T0402142

Advances in Computational Management Science 10

Editors:
H.M. Amman, Eindhoven, The Netherlands
B. Rustem, London, UK

For further volumes:
http://www.springer.com/series/5594

Wolfram Wiesemann

Optimization of Temporal Networks under Uncertainty

Wolfram Wiesemann
Imperial College London
Department of Computing
180 Queen's Gate
London, SW7 2AZ
United Kingdom
wwiesema@imperial.ac.uk

ISSN 1388-4301
ISBN 978-3-642-23426-2 e-ISBN 978-3-642-23427-9
DOI 10.1007/978-3-642-23427-9
Springer Heidelberg Dordrecht London New York

Library of Congress Control Number: 2011941397

Printed on acid-free paper

Springer is part of Springer Science+Business Media (www.springer.com)

Preface

A wide variety of decision problems in operations research are defined on temporal networks, that is, workflows of time-consuming tasks whose processing order is constrained by precedence relations. For example, temporal networks are used to formalize the management of projects, the execution of computer applications, the design of digital circuits and the scheduling of production processes. Optimization problems arise in temporal networks when a decision maker wishes to determine a temporal arrangement of the tasks and/or a resource assignment that optimizes some network characteristic such as the network's makespan (i.e., the time required to complete all tasks) or its net present value.

Optimization problems in temporal networks have been investigated intensively since the early days of operations research. To date, the majority of contributions focus on deterministic formulations where all problem parameters are known. This is surprising since parameters such as the task durations, the network structure, the availability of resources and the cash flows are typically unknown at the time the decision problem arises. The tacit understanding in the literature is that the decision maker replaces these uncertain parameters with their most likely or expected values to obtain a deterministic optimization problem. It is well documented in theory and practice that this approach can lead to severely suboptimal decisions.

The objective of this monograph is to survey state-of-the-art solution techniques for optimization problems in temporal networks that explicitly account for parameter uncertainty. Apart from theoretical and computational challenges, a key difficulty is that the decision maker may not be aware of the precise nature of the uncertainty. We therefore study several formulations, each of which requires different information about the probability distribution of the uncertain problem parameters. We discuss models that maximize the network's net present value, problems that minimize the network's makespan and multi-objective formulations that account for the costs associated with the temporal network. Throughout this book, emphasis is placed on tractable techniques that scale to industrial-size problems.

Many people have – directly or indirectly – contributed to the completion of this book. I would like to express my deepest gratitude to Professor Berç Rustem and Dr. Daniel Kuhn from the Department of Computing at Imperial College London. I have benefited greatly from their invaluable suggestions, enlightening advice and constant encouragement, both personally and scientifically. My sincere appreciation is extended to my colleague and valued friend Dr. Ronald Hochreiter. I would also like to thank Professor Wolfgang Domschke (Technische Universität Darmstadt) and Professor Robert Klein (Universität Augsburg) who whet my interest in optimization and decision making under uncertainty. Finally, I am deeply indebted to my family and Sharon.

United Kingdom *Wolfram Wiesemann*

Contents

List of Figures

List of Tables

Chapter 1
Introduction

1.1 Motivation

We define a *temporal network* as a directed, acyclic graph $G = (V, E)$ whose nodes $V = \{1, \ldots, n\}$ represent the network tasks and whose arcs $E \subseteq V \times V$ describe the temporal precedences between the tasks. This convention is known as *activity-on-node* notation; an alternative *activity-on-arc* notation is discussed in [DH02]. In our notation, an arc $(i, j) \in E$ signalizes that task j must not be started before task i has been completed. For ease of exposition, we assume that $1 \in V$ represents the unique source and $n \in V$ the unique sink of the network. This can always be achieved by introducing dummy nodes and/or arcs. We assume that the processing of each task requires a nonnegative amount of time. Depending on the problem under consideration, the tasks may also give rise to cash flows. Positive cash flows denote cash inflows (e.g., received payments), whereas negative cash flows represent cash outflows (e.g., accrued costs). Figure 1.1 illustrates a temporal network with cash flows.

Optimization problems arise in temporal networks when the decision maker is able to influence the processing of the network tasks. Most frequently, it is assumed that this is possible in one or two complementary ways. On one hand, the decision maker may be able to decide on the temporal orchestration of the tasks, that is, on the times at which the tasks are processed. On the other hand, the decision maker may be able to change the task durations through the assignment of resources. A rational decision maker influences the processing of the network tasks in order to optimize an objective function. In this book we consider several prominent objectives, namely the minimization of the network's makespan (i.e., the time required to process all tasks), the maximization of the network's net present value, and the minimization of the costs associated with the network. Other objectives (e.g., a level resource consumption or the lateness with respect to a deadline) are discussed in [DH02].

Temporal networks and their associated optimization problems are ubiquitous in operations research. In the following, we provide some illustrative examples.

W. Wiesemann, *Optimization of Temporal Networks under Uncertainty*,
Advances in Computational Management Science 10,
DOI 10.1007/978-3-642-23427-9_1, © Springer-Verlag Berlin Heidelberg 2012

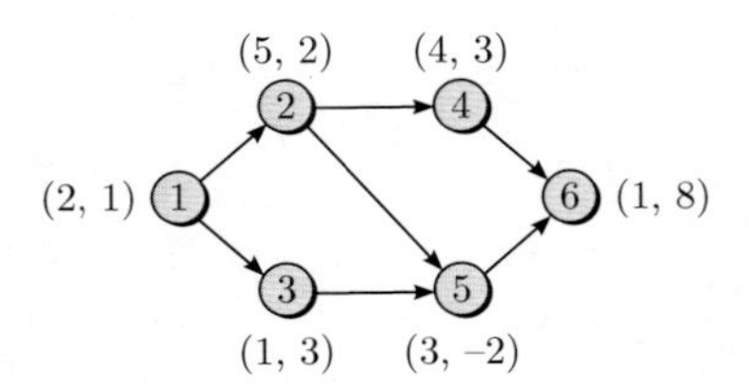

Fig. 1.1 Example temporal network. Attached to each node is the duration (first value) and the cash flow (second value) associated with the task

1. *Project scheduling*: Much of the research on temporal networks originates from the area of project scheduling, see [BDM+99, DH02, NSZ03, Sch05]. In project scheduling, the network tasks represent the various activities in a project (e.g., "conduct market research" or "develop prototype"), and the precedence relations describe temporal constraints between the activities (e.g., "the prototype cannot be developed before the market research has been completed"). The minimization of a project's makespan and the maximization of a project's net present value are among the most wide-spread objective functions in project scheduling. We will consider a project scheduling problem in Chap. 5.
2. *Execution of computer applications*: Computer applications can be described through flowgraphs whose nodes represent the application components and whose arcs describe the execution flow. Although a flowgraph typically accommodates non-deterministic flow constructs such as loops and conditional branches, it can be converted into a set of alternative execution flows, each of which constitutes a temporal network [vdAtHKB03, WHK08]. The execution of computer applications poses several challenging problems such as the scheduling of multiple applications on one or more processors and the assignment of resources (e.g., processor time, memory space and I/O access) to application commands, see [BEP+96]. The minimization of an application's runtime can be cast as a makespan minimization problem [WHK08]. We will discuss an application of temporal networks to service-oriented computing in Chap. 4.
3. *Design of digital circuits*: Modern VLSI (Very-Large-Scale Integration) circuits can contain millions of interconnected logical gates. A fundamental problem in VLSI design relates to the selection of suitable gate sizes [BKPH05]. The gate sizes crucially affect the three primary design objectives "operating speed", "total circuit size" and "power consumption". A circuit can be expressed as a temporal network whose tasks represent the gates and whose precedences denote the interconnections between the gates. Since the gate delay (i.e., the "task duration") is a function of the gate size, the maximization of the circuit speed, subject to constraints on the power consumption and the overall circuit size, can be cast as a makespan minimization problem in a temporal network. We will investigate circuit sizing problems in Chap. 6.
4. *Process scheduling*: A typical problem in process scheduling is to manufacture a set of products through a sequence of processing steps. Each processing step can be executed by a number of machines. At any time, a machine can process at most one product, and a product can be processed by at most one machine. Additionally, the processing times can depend on the assignment of resources

(e.g., fuel, catalysts and additional manpower). A common objective is to find a resource allocation and processing sequences that optimize the makespan or net present value of the production plan. Process scheduling problems are reviewed in [Bru07, Pin08].

In the remainder of this section, we highlight some of the difficulties that arise when the problem parameters of a temporal network are uncertain. To this end, let us first assume that all parameters are deterministic and that the resource assignment is fixed. We want to determine a vector of start times for the network tasks that optimizes the network's makespan or its net present value. The makespan is minimized by the following model:

$$\begin{aligned} &\underset{y}{\text{minimize}} && y_n + d_n \\ &\text{subject to} && y \in Y, \\ &\text{where} && Y = \left\{ y \in \mathbb{R}^n_+ \, : \, y_j \geq y_i + d_i \;\; \forall \, (i,j) \in E \right\}. \end{aligned} \tag{1.1}$$

In this problem, y_i and d_i denote the start time (a variable) and the duration (a parameter) of the ith task, respectively. The set Y contains the admissible start time vectors for the network tasks, that is, all start time vectors that satisfy the precedence constraints. Since n is the unique sink of the network, $y_n + d_n$ represents the network's makespan. Note that (1.1) constitutes a linear program that can be solved efficiently. Indeed, the minimal makespan can be determined much more efficiently if we exploit the following observation. Every admissible start time schedule $y \in Y$ has to satisfy $y_1 \geq 0$ and

$$y_j \geq \max_{i \in V} \{ y_i + d_i \, : \, (i,j) \in E \} \quad \text{for all } j \in V \setminus \{1\}.$$

Since the makespan is a non-decreasing function of y, the *early start schedule* $y^* \in Y$ with

$$y^*_j = \begin{cases} 0 & \text{if } j = 1, \\ \max_{i \in V} \left\{ y^*_i + d_i \, : \, (i,j) \in E \right\} & \text{otherwise} \end{cases} \tag{1.2}$$

is optimal. Note that the recursion is well-defined because G is acyclic. Hence, we can determine the minimal makespan through a topological sort. In Fig. 1.1, the minimal makespan of 12 time units is attained by the start time vector $y^* = (0, 2, 2, 7, 7, 11)^\top$.

The optimality of the early start schedule distinguishes the makespan from other objective functions in temporal networks. To illustrate this point, consider the following net present value maximization problem:

$$\begin{aligned} &\underset{y}{\text{maximize}} && \sum_{i \in V} \zeta_i \beta^{y_i} \\ &\text{subject to} && y \in Y. \end{aligned} \tag{1.3}$$

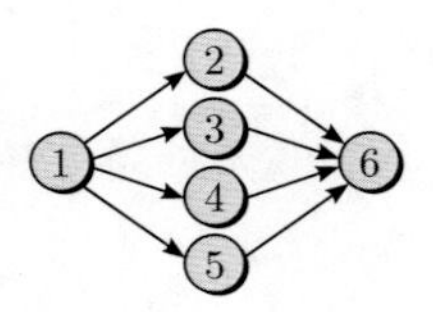

Fig. 1.2 Nominal models underestimate the makespan. In this temporal network, tasks 1 and n have duration zero, while the durations of the other tasks follow independent uniform distributions with support [0, 1]

Here, ζ_i denotes the cash flow arising at the start time y_i of task i, $\beta \in (0,1)$ represents the discount factor, and the set Y of admissible start time schedules is defined in (1.1). Although the objective function of (1.3) is nonconvex, the problem can be converted into an equivalent linear program by substituting the expressions β^{y_i} with new variables z_i, $i \in V$. We will elaborate on this substitution in Chap. 3. Note that problem (1.3) is no longer guaranteed to be optimized by the early start schedule y^* if negative cash flows are present. Indeed, for sufficiently large β, the net present value of the network in Fig. 1.1 is maximized by the start time vector $y = (0, 2, 2, 7, 8, 11)^\top$. As we will see throughout this book, the optimality of the early start schedule can dramatically simplify decision-making in temporal networks when uncertainty is present.

From the previous discussion we conclude that the makespan and the net present value of a deterministic temporal network can be optimized efficiently if the resource assignment is fixed. Let us now assume that the task durations are uncertain. A common suggestion is to solve a *nominal problem* where the uncertain task durations are replaced with their expected values. To see why this approach can be problematic, consider the temporal network $G = (V, E)$ with $V = \{1, \ldots, n\}$ and $E = \{(1, i) : 1 < i < n\} \cup \{(i, n) : 1 < i < n\}$, $n \geq 3$ [Elm05, MÖ1]. We illustrate the temporal network for $n = 6$ in Fig. 1.2. Assume that the tasks 1 and n have zero duration, while the durations d_i of the tasks $i \in \{2, \ldots, n-1\}$ follow independent uniform distributions with support [0, 1]. In this case, the expected duration of tasks 1 and n is zero, while all other tasks have an expected duration of $1/2$. From our previous discussion we know that the early start schedule $y^* = (0, \ldots, 0, 1/2)^\top$ minimizes the nominal makespan minimization problem, and hence the obtained estimate for the network's makespan is $1/2$. However, the probability that the makespan of the early start schedule does not exceed $t \in [0, 1]$ is given by the expression

$$\begin{aligned}\mathbb{P}(\max\{d_2, \ldots, d_{n-1}\} \leq t) &= \mathbb{P}(d_2 \leq t, \ \ldots, \ d_{n-1} \leq t) \\ &= \prod_{1<i<n} \mathbb{P}(d_i \leq t) \ = \ t^{n-2}.\end{aligned}$$

Thus, the probability to complete all tasks before time $t < 1$ goes to zero as n tends to infinity, and the approximation obtained from solving the nominal problem becomes increasingly weak.

More generally, one can show that the nominal problem always underestimates the expected makespan of the early start schedule. To see this, assume that the task durations d_i, $i \in V$, are random and that each task is started according to the early start policy y^*. Note that y^* constitutes a random vector now because it depends on the random task durations through (1.2). As before, the makespan is $y_n^* + d_n$. The right-hand side of (1.2) is convex and non-decreasing in y^* and d. Hence, we can reformulate the makespan as a convex function of the random task durations d_i by recursively replacing each component of y^* with its definition in (1.2). Jensen's inequality tells us that for a measurable convex function φ and a random vector d, $\varphi(\mathbb{E}(d)) \leq \mathbb{E}(\varphi(d))$. When we solve the nominal problem, we evaluate the left-hand side of this equation (deterministic makespan using expected durations) to approximate the right-hand side (expected makespan using random durations).

These rather pessimistic results on the approximation quality of nominal problems suggest that we should explicitly account for the stochastic nature of temporal networks. Unfortunately, the existence of precedence constraints severely complicates this goal. To see this, consider again the temporal network in Fig. 1.1 and assume that d_i, the duration of task $i \in V$, is described by its probability density function f_i and its cumulative distribution function F_i. For ease of exposition, we assume that the task durations are independently distributed. Following the exposition in [DH02], we want to determine the cumulative distribution function of the makespan if each task is started according to the early start schedule y^*. If we denote the cumulative distribution function of y_i^* by G_i, we obtain for the first three tasks

$$G_1(t) = \begin{cases} 1 & \text{if } t \geq 0, \\ 0 & \text{otherwise;} \end{cases} \qquad \text{and} \qquad G_2(t) = G_3(t) = \mathbb{P}(d_1 \leq t) = F_1(t).$$

The distribution of y_4 is obtained as follows:

$$G_4(t) = \mathbb{P}(d_1 + d_2 \leq t) = (F_1 * f_2)\,(t),$$

where $(\chi_1 * \chi_2)(t) := \int_{\tau \in \mathbb{R}} \chi_1(\tau)\, \chi_2(t - \tau)\, \mathrm{d}\tau$ denotes the convolution of two functions χ_1 and χ_2. The start time of task 5 depends on the maximum of two independent random variables:

$$\begin{aligned} G_5(t) &= \mathbb{P}(\max\{d_1 + d_2,\ d_1 + d_3\} \leq t) = \mathbb{P}(d_1 + \max\{d_2, d_3\} \leq t) \\ &= (f_1 * \widetilde{G})(t) = (f_1 * [F_2 \cdot F_3])\,(t), \end{aligned}$$

where

$$\begin{aligned} \widetilde{G}(t) &:= \mathbb{P}(\max\{d_2, d_3\} \leq t) = \mathbb{P}(d_2 \leq t,\ d_3 \leq t) \\ &= \mathbb{P}(d_2 \leq t)\, \mathbb{P}(d_3 \leq t) = F_2(t)\, F_3(t). \end{aligned}$$

Here, we used the notation $(\chi_1 \cdot \chi_2)(t) := \chi_1(t)\,\chi_2(t)$. Calculating G_6 is more involved as it depends on the maximum of *dependent* random variables:

$$\begin{aligned} G_6(t) &= \mathbb{P}(\max\{d_1 + d_2 + d_4,\ d_1 + d_2 + d_5,\ d_1 + d_3 + d_5\} \leq t) \\ &= \mathbb{P}(d_1 + \max\{d_2 + d_4,\ d_2 + d_5,\ d_3 + d_5\} \leq t) \\ &= (f_1 * \widehat{G})(t), \end{aligned}$$

where

$$\begin{aligned} \widehat{G}(t) &:= \mathbb{P}(\max\{d_2 + d_4,\ d_2 + d_5,\ d_3 + d_5\} \leq t) \\ &= \int_{\delta_2, \delta_5 \geq 0} \mathbb{P}(\delta_2 + d_4 \leq t)\mathbb{P}(\delta_2 + \delta_5 \leq t)\mathbb{P}(d_3 + \delta_5 \leq t)\, f_2(\delta_2)\ f_5(\delta_5)\, \mathrm{d}\delta_2\, \mathrm{d}\delta_5 \\ &= \int_{\substack{\delta_2, \delta_5 \geq 0, \\ \delta_2 + \delta_5 \leq t}} \mathbb{P}(d_4 \leq t - \delta_2)\,\mathbb{P}(d_3 \leq t - \delta_5)\ f_2(\delta_2)\ f_5(\delta_5)\, \mathrm{d}\delta_2\, \mathrm{d}\delta_5 \\ &= \int_{\substack{\delta_2, \delta_5 \geq 0, \\ \delta_2 + \delta_5 \leq t}} F_4(t - \delta_2)\ F_3(t - \delta_5)\ f_2(\delta_2)\ f_5(\delta_5)\, \mathrm{d}\delta_2\, \mathrm{d}\delta_5. \end{aligned}$$

The cumulative distribution function of the network's makespan is given by $G_6 * f_6$. Clearly, this approach becomes impractical for large networks. In fact, we cannot expect that there is an algorithm that determines the cumulative distribution function of the makespan efficiently. It has been shown in [Hag88] that even if the task durations are independent random variables with a two-valued support, the calculation of the expected value or any pre-specified quantile of the makespan of y^* is #PSPACE-hard. The situation is complicated by the practical difficulty to estimate the distributions of all task durations.

1.2 Book Outline

This book surveys solution techniques for optimization problems in temporal networks under uncertainty. The problems that we consider vary in the required information about the uncertain problem parameters (support, moments, probability distribution), the employed risk measure (unconditional and conditional expected values, quantiles and the worst case) and the objective function (makespan, net present value and costs). We apply our techniques to problems in project scheduling, computing and VLSI design. However, we stress that the proposed techniques apply to other application areas of temporal networks as well.

Apart from a review of some required background theory in Chap. 2, this book is divided into four chapters. Each of these chapters investigates a specific class of optimization problems in temporal networks.

In Chap. 3 we consider the maximization of a network's expected net present value when the task durations and cash flows are described by a discrete set of alternative scenarios with associated occurrence probabilities. In this setting, the choice of scenario-independent task start times frequently leads to infeasible schedules or severe losses in revenues. We present a technique that determines an optimal target processing time policy for the network tasks instead. Such a policy prescribes a task to be started as early as possible in the realized scenario, but never before its (scenario-independent) target processing time. We formulate the resulting model as a global optimization problem and describe a branch-and-bound algorithm for its solution. The solution scheme was first presented in the recent publication [WKR10].

We continue with a scenario-based description of the uncertain parameters in Chap. 4. Contrary to the expected net present value, however, we minimize the conditional value-at-risk (CVaR) of the network's makespan and the costs associated with the network. CVaR is a popular risk measure in decision theory which quantifies the expected shortfall below some percentile of a loss distribution. We apply the approach to the service composition problem in service-oriented computing. Given a formal specification of the desired behavior of a computer application, the service composition problem asks for a combination of software components that implements the application's behavior and at the same time maximizes the user's benefit. We quantify the user's benefit through several quality-of-service parameters, namely the application's runtime, the service invocation costs, as well as the availability and reliability of the selected services. We compare this model with a nominal problem formulation from the literature, and we show that the nominal model can lead to overly risky decisions. The service composition model was first described in [WHK08].

Chapter 5 investigates a resource allocation model that minimizes the makespan of a temporal network. The model accommodates multiple resources and decision-dependent task durations that are inspired by microeconomic theory. In the first part of the chapter, we elaborate a deterministic problem formulation. In the second part, we enhance the model to account for uncertain problem parameters. Assuming that the first and second moments of these parameters are known, the stochastic model minimizes an approximation of the value-at-risk (i.e., a specified quantile) of the network's makespan. As a salient feature, the described approach employs a scenario-free formulation which approximates the durations of the network's task paths via normal distributions. We discuss an extension of the model to situations in which the moments of the random parameters are ambiguous, and we describe an iterative solution procedure. The resource allocation model discussed in this chapter is based on the paper [WKRa].

In Chap. 6 we study a robust resource allocation problem in temporal networks where the task durations are uncertain, and the goal is to minimize the worst-case makespan. We show that this problem is generically $\mathcal{NP}$-hard. We then discuss families of optimization problems that provide convergent lower and upper bounds on the optimal value of the problem. The upper bounds correspond to feasible allocations whose objective values are bracketed by the bounds. Hence, we obtain a series of feasible allocations that converge to the optimal solution and whose optimality gaps can be quantified. We study the scalability of the solution approach on random test instances, and we discuss a case study in VLSI design. The bounding approach was first published in [WKRb].

1.3 Notation

By default, all vectors are column vectors. We denote the p-norm of a vector x by $\|x\|_p$. We denote by e_k the kth canonical basis vector, while e denotes the vector whose components are all ones. In both cases, the dimension will usually be clear from the context. We denote the set of real numbers, nonnegative real numbers and strictly positive real numbers by $\mathbb{R}$, $\mathbb{R}_+$ and $\mathbb{R}_{++}$, respectively. We denote the set of natural numbers (including zero) by $\mathbb{N}$ ($\mathbb{N}_0$).

We say that a set has a *tractable representation* if set membership can be described by finitely many convex constraints and, potentially, auxiliary variables. Similarly, a function has a tractable representation if its epigraph does. An *explicit* optimization problem has finitely many variables and constraints.

Some of the chapters in this book require additional notation. We defer the introduction of that notation to the relevant chapters.

Chapter 2
Background Theory

We start with a review of deterministic optimization problems in temporal networks. We then discuss three popular methodologies to model and solve generic optimization problems under uncertainty. We close with an overview of the issues that arise when these methodologies are applied to temporal networks, and we provide a survey of the relevant literature. More specific reviews of related work are provided in the Chaps. 3–6.

2.1 Temporal Networks

The literature on temporal networks is vast and has been reviewed, amongst others, in [BDM$^+$99, BEP$^+$96, BKPH05, Bru07, DH02, FL04, NSZ03, Pin08, Sch05]. Instead of giving a detailed account of all contributions, we classify some of the most popular research directions according to the three dimensions "resources", "network" and "objective". More elaborate classification schemes can be found in [BDM$^+$99, Bru07, DH02].

Resource characteristics: Optimization problems in temporal networks may assume that a resource allocation has been fixed, or they can involve the assignment of one or multiple resources. In the latter case, we can distinguish between three prevalent types of resources. *Non-renewable resources* are available in pre-specified quantities and are not replenished during the planning horizon. Typical examples of non-renewable resources are capital and man-hours. In contrast, *renewable resources* are replenished every time period, but the decision maker has to meet specified per-period consumption quotas. Examples of renewable resources are processing times on manufacturing machines and processors. In practice, many resources are *doubly constrained*, that is, they share the restrictions of non-renewable and renewable resources. Other resource characteristics include time windows during which the resources are available, as well as spatial aspects (e.g., immobile resources such as a shipyard).

W. Wiesemann, *Optimization of Temporal Networks under Uncertainty*, Advances in Computational Management Science 10,
DOI 10.1007/978-3-642-23427-9_2, © Springer-Verlag Berlin Heidelberg 2012

Network characteristics: Network characteristics describe the properties of the network tasks and precedences. Tasks are *preemptive* if their processing can be interrupted to execute other tasks. For example, modern operating systems use preemptive multitasking to generate the illusion of executing multiple computer applications in parallel on a single processor. If the execution of network tasks must not be interrupted, then the tasks are called *non-preemptive*. Project scheduling, circuit design and many problems in process scheduling assume that the network tasks are non-preemptive. In the introduction, we assumed that all precedences in the temporal network are of *finish-start* type, that is, an arc from node i to node j in the temporal network prescribes that task j cannot be started before task i has been completed. Alternatively, one can consider *generalized precedences* that stipulate lower and upper bounds on the time that may pass between the start and completion of any two network tasks. Other network characteristics include time windows during which the tasks must be executed (e.g., ready times and deadlines) and cash flows that arise when the tasks are processed.

Objective function: One commonly distinguishes between *regular objective functions*, which are optimized by the early start schedule (1.2), and *nonregular objective functions*, which may not be optimized by the early start schedule. Typical regular objective functions are the makespan and the lateness of the makespan beyond a given deadline. Examples of nonregular objectives are the net present value and a level resource consumption.

The methods studied in this book are primarily applicable to temporal networks with non-preemptive tasks and non-renewable resources. Chapter 3 assumes that the resource allocation is fixed and maximizes the expected net present value under generalized precedences. In Chap. 4–6 we determine assignments of non-renewable resources under finish-start precedences. Chapter 4 studies a multi-objective problem that considers discrete resources (web services) and optimizes the conditional value-at-risk of the makespan and resource costs. Chapters 5 and 6 assume continuous resources (e.g., capital or man-hours). Chapter 5 optimizes quantiles of the makespan, whereas Chap. 6 minimizes the worst-case makespan.

2.2 Optimization Under Uncertainty

In practice, most managerial decisions are taken under significant uncertainty about relevant data such as future market developments and resource availabilities. If such decision problems are formulated as optimization models, the models contain parameters whose values are uncertain. In the following, we review three popular approaches to model and solve optimization problems with uncertain parameters. In the remainder of the book, we will present applications of two of these approaches to optimization problems in temporal networks.

Fig. 2.1 Temporal structure of a two-stage (*left*) and a multi-stage (*right*) recourse problem. In the *left time line*, the wait-and-see decision y may depend on x and ξ. In the *right time line*, the wait-and-see decision y^t may depend on x and ξ^s, $s < t$

2.2.1 Stochastic Programming

Stochastic programming models the uncertain problem parameters as random variables with known probability distributions. One of the basic models is the *two-stage recourse problem*.

$$\inf_{x \in X} \{f(x) + \mathbb{E}[Q(x;\xi)]\}, \tag{2.1a}$$

where

$$Q(x;\xi) = \inf_{y \in Y(x,\xi)} \{q(y;x,\xi)\}. \tag{2.1b}$$

In this problem, the parameter vector ξ is assumed to be uncertain. The decision maker needs to take a *here-and-now decision* $x \in X$ before the value of ξ is known, while the *wait-and-see decision* $y \in Y(x,\xi)$ can be selected under full knowledge of ξ. Conceptually, we can assume that x is chosen at the beginning of time period 1, ξ is revealed during time period 1, and y is selected at the beginning of time period 2 (after ξ is known), see Fig. 2.1, left. The goal is to minimize the sum of deterministic first-stage costs $f(x)$ and expected second-stage costs $\mathbb{E}[Q(x;\xi)]$, where the expectation is taken with respect to ξ. Note that for any value of x and ξ, the second-stage problem $Q(x;\xi)$ is deterministic. If there is a finite set of values $\xi^1, \xi^2, \ldots$ such that $\xi \in \{\xi^1, \xi^2, \ldots\}$ with probability one, then (2.1) can be formulated as an explicit optimization problem. Otherwise, (2.1) can be approximated by a surrogate model that replaces the probability distribution of ξ with a finite-valued approximation. In either case, the resulting optimization model has the structure of a *scenario fan* whose branches represent the possible realizations of ξ, see Fig. 2.2, left. The decision maker's information set (i.e., the set of scenarios that may be realized) is shown in Fig. 2.2, right. At the beginning of the first time period, the decision maker is unaware of the realized scenario ξ^k. The information set therefore contains all scenarios. In the second time period, the decision maker knows the realized scenario ξ^k. The information set has therefore shrunk to one of the singleton sets $\{\xi_1\}, \ldots, \{\xi_5\}$.

The use of the expected value in (2.1a) reflects the assumption that the decision maker is *risk-neutral*. In many applications of temporal networks, such as project management and production scheduling, this assumption may not hold. Instead,

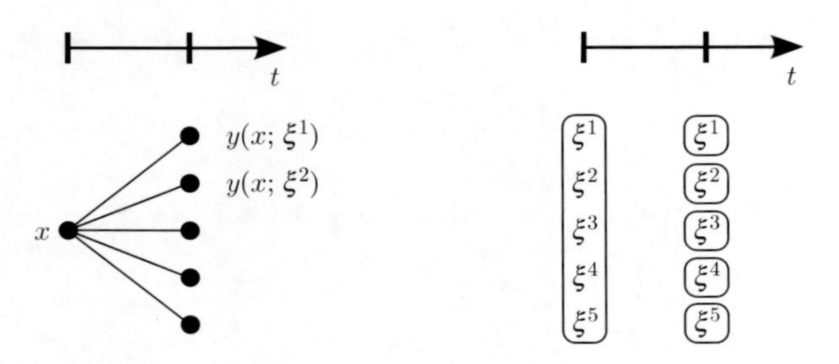

Fig. 2.2 Scenario representation of two-stage recourse problems. The *left chart* shows that for each realization (scenario) ξ^k of the random vector ξ, a separate recourse decision $y(x;\xi^k)$ can be selected. The *right chart* visualizes the acquisition of information over time

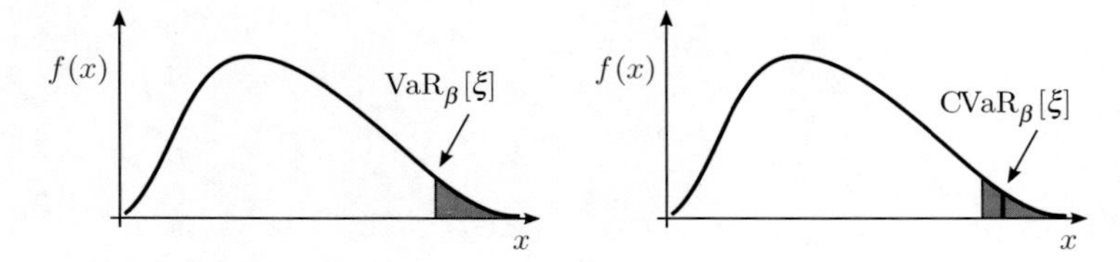

Fig. 2.3 Value-at-risk (*left*) and conditional value-at-risk (*right*) of a continuous random variable ξ with probability density function $f(x)$

the decision maker is *risk-averse* and prefers solutions that do not just perform well "on average", but that also perform satisfactory "in most cases". The most widely used approach to obtain risk-averse solutions is to minimize the variance of $Q(x;\xi)$, which traces back to the seminal paper [Mar52] on financial portfolio selection. However, minimizing the variance of $Q(x;\xi)$ penalizes both the excess and the shortfall of $Q(x;\xi)$ with respect to its expected value $\mathbb{E}[Q(x;\xi)]$. This may not be appropriate for optimization problems in temporal networks. If the goal is to minimize the makespan, for example, a decision maker only wants to penalize upward deviations from the expected value (i.e., delays), whereas downward deviations are indeed desirable. A decision maker may therefore prefer to optimize a one-sided quantile-based risk measure such as the value-at-risk (VaR):

$$\mathrm{VaR}_\beta[Q(x;\xi)] = \inf\{\alpha \in \mathbb{R} : \mathbb{P}(Q(x;\xi) > \alpha) \leq 1 - \beta\}.$$

For ease of exposition, we assume for the remainder of this section that the random parameters ξ follow a continuous probability distribution. For $\beta \in [0, 1]$, we can then interpret the β-VaR of $Q(x;\xi)$ as the β-quantile of $Q(x;\xi)$. For high values of β (e.g., $\beta \geq 0.9$), minimizing the β-VaR of $Q(x;\xi)$ favors solutions that perform well in most cases. The VaR of a random variable ξ is illustrated in Fig. 2.3, left.

In recent years, VaR has come under criticism due to its nonconvexity, which makes the resulting optimization models difficult to solve. Moreover, the nonconvexity implies that VaR is not sub-additive and hence not a coherent risk measure in

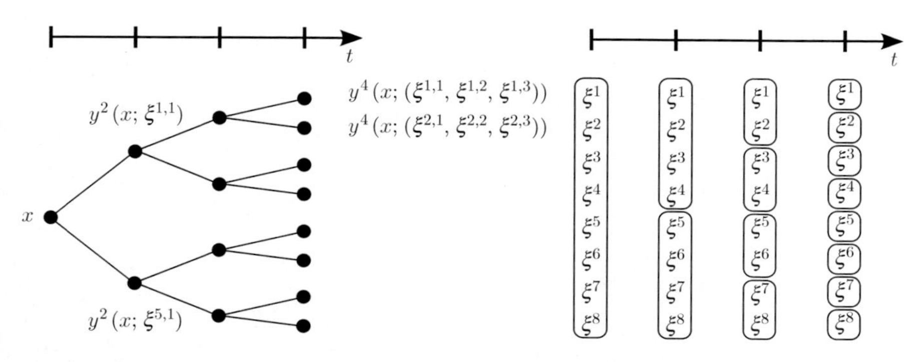

Fig. 2.4 Scenario representation of multi-stage recourse problems. In analogy to the scenario fan in Fig. 2.2, left, the *left chart* visualizes the scenario tree associated with a multi-stage recourse problem, whereas the *right chart* shows the acquisition of information over time

the sense of [ADEH99]. Both drawbacks are rectified by the conditional value-at-risk (CVaR), which is defined as follows:

$$\text{CVaR}_\beta\left[Q(x;\xi)\right] = \mathbb{E}\left(Q(x;\xi) \mid Q(x;\xi) \geq \text{VaR}_\beta\left[Q(x;\xi)\right]\right).$$

The β-CVaR of $Q(x;\xi)$ represents the expected value of $Q(x;\xi)$ under the assumption that $Q(x;\xi)$ exceeds its β-VaR, that is, under the assumption that $Q(x;\xi)$ is among the $(1-\beta)\cdot 100\%$ "worst" outcomes. By definition, the β-CVaR exceeds the β-VaR for any $\beta \in [0,1]$. The CVaR of a random variable ξ is illustrated in Fig. 2.3, right. It has been shown in [RU00] that the β-CVaR is equivalent to

$$\inf_{\alpha\in\mathbb{R}} \left\{ \alpha + \frac{1}{1-\beta}\mathbb{E}\left[Q(x;\xi) - \alpha\right]^+ \right\},$$

where $[x]^+ = \max\{x, 0\}$. Hence, the techniques presented for recourse problems with expected value objective functions are directly applicable to optimization problems involving CVaR.

So far, all of the models assumed a two-stage structure (decision – realization of uncertainty – decision). In a *multi-stage recourse problem,* the parameter vector ξ can be subdivided into vectors $\xi^1, \dots, \xi^T$ such that $\xi = (\xi^1, \dots, \xi^T)$ and ξ^t is revealed during time period $t = 1, \dots, T$. The decision maker can take a recourse decision y^t at the beginning of every time period $t = 2, \dots, T+1$, and y^t may depend on the values of $\xi^1, \dots, \xi^{t-1}$, see Fig. 2.1, right. Note that y^t may *not* depend on the values of ξ^s, $s \geq t$, since this information is not available at the time the recourse decision y^t is taken. This causality requirement is called *non-anticipativity*. If the probability distribution of ξ has finitely many values, then the optimization model associated with a multi-stage recourse problem has the structure of a *scenario tree,* see Fig. 2.4. In the left chart of that figure, $\xi^{k,t}$ denotes the tth subvector of the scenario $\xi^k = (\xi^{k,1}, \dots, \xi^{k,T})$. Each path from the root node to a leaf node constitutes one scenario. Two scenarios ξ^k and ξ^l are undistinguishable at the beginning of period t if $\xi^{k,s} = \xi^{l,s}$ for all $s < t$. In this case, ξ^k and ξ^l are contained in the same information set at time t, and non-anticipativity stipulates that $y^t(x; (\xi^{k,1}, \ldots, \xi^{k,t-1})) = y^t(x; (\xi^{l,1}, \ldots, \xi^{l,t-1}))$. For example, non-anticipativity requires that $y^2(x; \xi^{k,1}) = y^2(x; \xi^{l,1})$ for $k, l \in \{1, \ldots, 4\}$ and $y^3(x; (\xi^{5,1}, \xi^{5,2})) = y^3(x; (\xi^{6,1}, \xi^{6,2}))$. In analogy to two-stage recourse problems, a multi-stage recourse problem can be approximated by a surrogate model that replaces the probability distribution of ξ with a finite-valued approximation if ξ can attain infinitely many values. While convex two-stage recourse problems can be approximated efficiently, multi-stage problems "generically are computationally intractable already when medium-accuracy solutions are sought" [SN05]. Multi-stage recourse problems therefore constitute very difficult optimization problems.

Apart from recourse problems, stochastic programming studies problems with *chance constraints*. The basic two-stage chance constrained problem can be formulated as follows:

$$\inf_{x \in X} \{ f(x) \; : \; \mathbb{P}(Q(x; \xi) \leq 0) \geq 1 - \epsilon \}, \tag{2.2}$$

where Q is defined in (2.1b). The temporal structure of problem (2.2) is the same as for two-stage recourse problems, see Fig. 2.1, left. The goal is to find a here-and-now decision x such that with a probability of at least $1 - \epsilon$, there is a wait-and-see decision $y(x; \xi) \in Y(x, \xi)$ that satisfies $q(y(x; \xi); x, \xi) \leq 0$. Note that problem (2.2) is equivalent to the following problem with a VaR constraint:

$$\inf_{x \in X} \{ f(x) \; : \; \mathrm{VaR}_{1-\epsilon} [Q(x, \xi)] \leq 0 \}. \tag{2.2'}$$

It is therefore not surprising that chance constrained problems inherit the computational difficulties of VaR optimization problems. Indeed, even if the second-stage problem Q is a linear program, the feasible region of (2.2) is typically nonconvex and disconnected [Pré95]. Moreover, calculating the left-hand side of the constraint in (2.2) requires the evaluation of a multi-dimensional integral, which itself constitutes a difficult problem. As a result, most solution approaches for (2.2) settle for approximate solutions. A popular approximation is obtained by replacing the $(1 - \epsilon)$-VaR in (2.2') with the $(1 - \epsilon)$-CVaR:

$$\inf_{x \in X} \{ f(x) \; : \; \mathrm{CVaR}_{1-\epsilon} [Q(x, \xi)] \leq 0 \}. \tag{2.2''}$$

Since CVaR represents an upper bound on VaR, this formulation provides a conservative approximation to problem (2.2'), that is, any $x \in X$ that is feasible in (2.2") is also feasible in (2.2') and (2.2). Similar to recourse problems, chance constrained problems can be extended to multiple decision stages.

For an in-depth treatment of stochastic programming, the reader is referred to the textbooks [KW94, Pré95, RS03]. We will consider two-stage recourse problems that optimize the expected value and CVaR in Chaps. 3 and 4, respectively. Chapter 5 studies an approximation of VaR that does not rely on scenario discretization.

2.2.2 *Robust Optimization*

In its basic form, robust optimization studies semi-infinite problems of the following type:

$$\inf_{x \in X} \{ f(x) \; : \; g_i(x;\xi) \leq 0 \;\; \forall \, \xi \in \Xi, \; i = 1, \ldots, I \} . \tag{2.3}$$

We interpret x as a here-and-now decision and ξ as an uncertain parameter vector with support Ξ. The goal is to minimize the deterministic costs $f(x)$ while satisfying the constraints for all possible realizations of ξ. Note that (2.3) is a single-stage problem since it does not contain any recourse decisions. If Ξ constitutes a finite set of scenarios $\xi^1, \xi^2, \ldots,$ then (2.3) can be formulated as an explicit optimization problem. If Ξ is of infinite cardinality, then (2.3) can be solved with iterative solution procedures from semi-infinite optimization [HK93, RR98]. One of the key contributions of robust optimization has been to show that for sets Ξ of infinite cardinality but specific structure, one can apply duality theory to transform problem (2.3) into an explicit optimization problem. We illustrate this approach with an example.

Example 2.2.1. Assume that $I = 1$, $X \subseteq \mathbb{R}^n$, $\Xi = \left\{ \xi \in \mathbb{R}^k_+ \; : \; W\xi \leq h \right\}$ for $W \in \mathbb{R}^{m \times k}$ and $h \in \mathbb{R}^m$, and $g_1(x;\xi) = \xi^\top A x$ for $A \in \mathbb{R}^{k \times n}$. Also assume that Ξ is nonempty and bounded. We can then reformulate the constraint in problem (2.3) as follows:

$$\begin{aligned}
g_1(x;\xi) \leq 0 \quad \forall \, \xi \in \Xi \quad & \Leftrightarrow \quad \sup_{\xi \in \Xi} \{ g_1(x;\xi) \} \leq 0 \\
& \Leftrightarrow \quad \max_{\xi \in \mathbb{R}^k_+} \left\{ \xi^\top A x \; : \; W\xi \leq h \right\} \leq 0 \\
& \Leftrightarrow \quad \min_{\lambda \in \mathbb{R}^m_+} \left\{ h^\top \lambda \; : \; W^\top \lambda \geq Ax \right\} \leq 0 \\
& \Leftrightarrow \quad h^\top \lambda \leq 0, \; W^\top \lambda \geq Ax \qquad \text{for some } \lambda \in \mathbb{R}^m_+ .
\end{aligned}$$

Here, the third equivalence follows from strong linear programming duality. We have thus transformed the semi-infinite constraint in problem (2.3) into a finite number of constraints that involve x and new auxiliary variables λ.

Much of the early work on robust optimization focuses on generalizations of the reformulation scheme illustrated in Example 2.2.1. Unfortunately, single-stage models such as (2.3) are too restrictive for decision problems in temporal networks. Indeed, the task start times can typically be chosen as a wait-and-see decision, and optimization problems that account for this flexibility provide significantly better solutions. We discuss this issue in more detail in Chaps. 3–6. We are therefore interested in *two-stage robust optimization problems* such as the following one:

$$\inf_{x \in X} \sup_{\xi \in \Xi} \inf_{y \in Y(x,\xi)} \{f(x) + q(y; x, \xi)\} . \tag{2.4}$$

Here, q is the objective function of the second-stage problem Q defined in (2.1b), and $Y(x, \xi) \subseteq \mathbb{R}^l$. In this problem, the here-and-now decision x is accompanied by a wait-and-see decision $y \in Y(x, \xi)$ that can be selected under full knowledge of ξ. The temporal structure of this problem is similar to the two-stage recourse problem (2.1), see Fig. 2.1, left. The goal is to minimize the sum of first-stage costs $f(x)$ and worst-case second-stage costs $\sup_{\xi \in \Xi} Q(x; \xi)$, see (2.1b), where the worst case is taken with respect to ξ. Two-stage robust optimization problems are generically intractable, see [BTGGN04]. A tractable approximation can be derived from the following identity.

Observation 2.2.1 *For any* $X \subseteq \mathbb{R}^n$, $\Xi \subseteq \mathbb{R}^k$ *and* $Y(x, \xi) \subseteq \mathbb{R}^l$, *we have*

$$\inf_{x \in X} \sup_{\xi \in \Xi} \inf_{y \in Y(x,\xi)} \{f(x) + q(y; x, \xi)\} \quad = \quad \inf_{\substack{x \in X, \\ y \in \mathcal{Y}(x)}} \sup_{\xi \in \Xi} \{f(x) + q(y(\xi); x, \xi)\} , \tag{2.5a}$$

where for $x \in X$,

$$\mathcal{Y}(x) = \left\{(y : \Xi \mapsto \mathbb{R}^l) \; : \; y(\xi) \in Y(x, \xi) \;\; \forall \, \xi \in \Xi\right\} . \tag{2.5b}$$

Observation 2.2.1 allows us to reduce the min–max–min problem (2.4) to the min–max problem on the right-hand side of (2.5a) at the cost of augmenting the set of first-stage decisions. For a given here-and-now decision $x \in X$, $\mathcal{Y}(x)$ denotes the space of all functions on Ξ that map parameter realizations to feasible wait-and-see decisions. A function y is called a *decision rule* because it specifies the second-stage decision in (2.4) as a function of the uncertain parameters ξ. Note that the choice of an appropriate decision rule on the right-hand side of (2.5a) is part of the first-stage decision. The identity (2.5a) holds regardless of the properties of X and Ξ because $\mathcal{Y}(x)$ does not impose any structure on the decision rules (such as measurability).

Since $\mathcal{Y}(x)$ constitutes a function space, further assumptions are required to ensure that the problem on the right-hand side of (2.5a) can be solved. A popular approach is to restrict $\mathcal{Y}(x)$ to the space of affine or piecewise affine functions of ξ, see [BTGN09, CSSZ08, KWG]. As we will show in Chap. 6, this restriction allows us to reformulate the model on the right-hand side of (2.5a) as an explicit optimization problem. Figure 2.5 compares the scenario approximation from the previous section with the decision rule approximation. In the left chart of that figure, the support Ξ of the random parameters ξ is replaced with a discrete-valued approximation. For each possible realization (scenario) ξ^k, an individual second-stage decision $y(x; \xi^k)$ may be chosen. In the right chart of Fig. 2.5, the

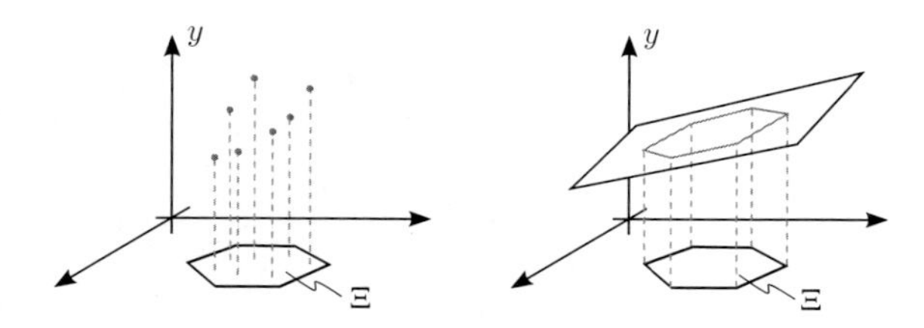

Fig. 2.5 Approximations employed by two-stage recourse problems (*left*) and two-stage robust optimization problems (*right*) for a random vector ξ with a continuous probability distribution

support Ξ of the random parameters ξ remains unchanged, but the second-stage decision $y(x;\xi)$ is restricted to be an affine function of ξ.

For an introduction to robust optimization, see [BS04, BTGN09]. Two-stage robust optimization problems are discussed in [BTGN09, CSSZ08, JLF07, LJF04, LLMS09, Sti09]. In recent years, the theory of robust optimization has been extended to recourse problems and chance constrained problems. For further details, see [BP05, BTGN09, CSST10, DY10, GS10]. In Chap. 6 we will solve a makespan minimization problem as a two-stage robust optimization problem. Instead of approximating the optimal second-stage decision via decision rules, however, this chapter presents a technique that provides convergent lower and upper bounds on the optimal value of the problem. The upper bounds correspond to feasible solutions whose objective values are bracketed by the bounds. We will compare that method with two popular classes of decision rules.

2.2.3 Stochastic Dynamic Programming

Stochastic dynamic programming studies the modeling and solution of optimization problems via *Markov decision processes* (MDPs). MDPs model dynamic decision problems in which the outcomes are partly random and partly under the control of the decision maker. At each time period, the MDP is in some state s, and the decision maker takes an action a. The state s' in the successive time period is random and depends on both the current state s and the selected action a. However, the new state does *not* depend on any other past states or actions: this is the *Markov property*. For each transition of the MDP, the decision maker receives a reward that depends on the old state, the new state and the action that triggered the transition.

In the following, we will restrict ourselves to discrete-time MDPs with finite state and action spaces. We therefore assume that an MDP is defined through its state space $\mathcal{S} = \{1,\ldots,S\}$, its action space $\mathcal{A} = \{1,\ldots,A\}$, and a discrete planning horizon $\mathcal{T} = \{0,1,2,\ldots\}$ that can be finite or infinite. The initial state is a random variable with known probability distribution p_0. If action $a \in \mathcal{A}$ is chosen in state $s \in \mathcal{S}$, then the subsequent state is $s' \in \mathcal{S}$ with probability $p(s'|s,a)$. We assume that the probabilities $p(s'|s,a)$, $s' \in \mathcal{S}$, sum up to one for each state–action pair $(s,a) \in \mathcal{S} \times \mathcal{A}$. The decision maker receives an expected reward of $r(s,a,s') \in \mathbb{R}$ if action $a \in \mathcal{A}$ is chosen in state $s \in \mathcal{S}$ and the subsequent state is $s' \in \mathcal{S}$. Without loss of generality, we can assume that every action is admissible in every state. Indeed, if action $a \in \mathcal{A}$ is not allowed in state $s \in \mathcal{S}$, then we can "forbid" this action by setting all rewards $r(s, a, s')$, $s' \in \mathcal{S}$, to a large negative value. Figure 2.6 visualizes the structure of an MDP.

At each stage, the MDP is controlled through a policy $\pi = (\pi_t)_{t \in \mathcal{T}}$, where $\pi_t(a|s_0, a_0, \dots, s_{t-1}, a_{t-1}; s_t)$ represents the probability to choose action $a \in \mathcal{A}$ if the current state is s_t and the state-action history is given by the vector $(s_0, a_0, \dots, s_{t-1}, a_{t-1})$. Note that contrary to the state transitions of the MDP, the policy π need not be Markovian. If the planning horizon $\mathcal{T}$ is infinite, then we can evaluate a policy π in view of its *expected total reward* under the discount factor $\lambda \in (0, 1)$:

$$\mathbb{E}\left[\sum_{t=0}^{\infty} \lambda^t r(s_t, a_t, s_{t+1}) \;\middle|\; s_0 \sim p_0\right]. \tag{2.6}$$

Here, $\mathbb{E}$ denotes the expectation with respect to the random process defined by the transition probabilities p and the policy π. The notation $s_0 \sim p_0$ indicates that the initial state s_0 is a random variable with probability distribution p_0. If the planning horizon $\mathcal{T}$ is finite, say $\mathcal{T} = \{0, 1, \dots, T\}$, then we can evaluate a policy π in view of its *expected total reward* without discounting:

$$\mathbb{E}\left[\sum_{t=0}^{T} r(s_t, a_t, s_{t+1}) \;\middle|\; s_0 \sim p_0\right]. \tag{2.7}$$

For a fixed policy π, the *policy evaluation problem* asks for the value of expression (2.6) or (2.7). The *policy improvement problem*, on the other hand, asks for a policy π that maximizes (2.6) or (2.7). For both objective functions, the policy evaluation and improvement problems can be solved efficiently via policy and value iteration.

Example 2.2.2 (Inventory Management). Consider the following infinite horizon inventory problem. At the beginning of each time period, the decision maker can order $a \in \mathbb{N}_0$ units of a product at unit costs c. The ordered products arrive at the beginning of the next period. During each period, an independent and identically distributed random demand δ arises for the product. This demand is served at a unit

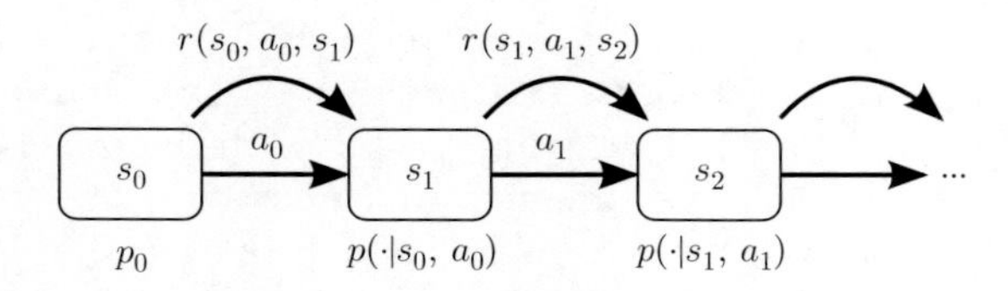

Fig. 2.6 Structure of a Markov decision process. The process starts in state $s_0 \in \mathcal{S}$, which follows the probability distribution p_0. After the action $a_0 \in \mathcal{A}$ is chosen, the new state $s_1 \in \mathcal{S}$ follows the probability distribution $p(\cdot|s_0, a_0)$, and an expected reward $r(s_0, a_0, s_1)$ is received

price p from the current inventory, and there is no backlogging (i.e., demand that cannot be satisfied within the period is lost). The inventory can hold at most I units of the product. The goal is to find an inventory control policy that maximizes the expected total reward under some discount factor λ.

We can formulate this problem as an infinite horizon MDP as follows. The state set $\mathcal{S} = \{0, \ldots, I\}$ describes the inventory level at the beginning of each time period. In state $s \in \mathcal{S}$, the admissible actions $\{0, \ldots, I - s\}$ determine the order quantity. Note that the actions are state-dependent in this example. The transition probabilities are

$$p(s'|s,a) = \begin{cases} \mathbb{P}(\delta = s + a - s') & \text{if } s' \neq 0, \\ \sum_{i=s+a}^{\infty} \mathbb{P}(\delta = i) & \text{otherwise,} \end{cases}$$

and the rewards are given by $r(s,a,s') = p(s + a - s') - ca$. Here we assume that the random demand δ is nonnegative with probability one. A policy π could order $\omega \in \mathbb{N}$ units whenever the current inventory falls below some threshold $\Omega \in \mathbb{N}_0$. This policy is defined through $\pi_t(a|s_0, a_0, \ldots, s_{t-1}, a_{t-1}; s_t) = 1$ if $s_t < \Omega$ and $a = \omega$, or $s_t \geq \Omega$ and $a = 0$, and $\pi_t(a|s_0, a_0, \ldots, s_{t-1}, a_{t-1}; s_t) = 0$ otherwise. Note that this policy π is Markovian.

There are numerous variations of the Markov decision process defined in this section. For an overview of the major models and solution approaches, see [Ber07, Put94].

In recent years, an approximation scheme called "approximate dynamic programming" has received much attention. The interested reader is referred to the textbooks [BT96, Pow07]. In this book, we will not consider the application of Markov decision processes to temporal networks. The reader is referred to [BR97, KA86, TSS06] and there references therein.

2.3 Optimization of Temporal Networks under Uncertainty

Decisions in temporal networks are often taken under significant uncertainty about the network structure (i.e., the tasks and precedences of the network), the task durations, the ready times and deadlines of the tasks, the cash flows and the availability of resources. In this book, we focus on problems in which the task durations (Chaps. 3–6), the cash flows (Chaps. 3 and 4), the network structure (Chap. 4), and the tasks' ready times and deadlines (Chap. 3) are uncertain. A problem that accounts for uncertain resource availabilities is described in [Yan05]. Further reviews on problems with uncertain network structure are provided in [Neu79, Neu99, Pri66].

An optimization problem under uncertainty needs to specify *when* information about the uncertain parameters becomes available, and *what* information is revealed about them. Both issues are straightforward in the optimization problems reviewed in Sect. 2.2. In a multi-stage recourse problem, for example, we observe the subvector ξ^t of the uncertain parameters ξ at the beginning of time period $t + 1$, see Fig. 2.1, right. Likewise, in a stochastic dynamic program, we observe the current state of the MDP at the beginning of each time period.

The situation is different for temporal networks, and it is this difference that complicates the modeling and solution of decision problems in temporal networks. It is customary to assume that the duration and cash flow of a task is observed when the task is completed. However, the completion time of a task depends on the task's start time, which is chosen by the decision maker. Hence, in contrast to the problems studied in Sect. 2.2, the times at which we learn about the random parameters depend on the chosen decision. Recourse problems with decision-dependent uncertainty are studied in [GG06, JWW98], and a robust optimization problem with decision-dependent uncertainty is formulated in [CGS07]. However, the resulting optimization problems are computationally demanding, and they typically have to undergo drastic simplifications before they can be solved.

Apart from the time points at which information becomes available, optimization problems in temporal networks differ from other problems in the type of the revealed information. In many cases, the task durations and cash flows in a temporal network do not correspond to individual parameters, but they are functions of multiple parameters (as is the case in factor models). In such problems, we do not observe the uncertain parameters themselves, but we accumulate knowledge about them with the completion of each task. We can use this information to exclude parameter realizations that are not compatible with the observed durations and cash flows. In contrast, the multi-stage recourse and robust optimization problems reviewed in Sect. 2.2 assume that the decision maker can directly observe the uncertain parameter vector ξ.

Similar to the problems in Sect. 2.2, optimization problems in temporal networks can contain here-and-now as well as wait-and-see decisions. Here-and-now decisions are taken before any of the network tasks are started, whereas a wait-and-see decision associated with task $i \in V$ (e.g., its start time or resource assignment) may depend on all information that is available at the time task i is started. Since the early start schedule optimizes regular objective functions (see Sect. 2.1), it is relatively straightforward to model the task start times as a wait-and-see decision in makespan minimization problems. We will consider problems with a here-and-now resource allocation and wait-and-see task start times in Chaps. 4–6. The situation is fundamentally different in net present value maximization problems where the early start schedule is no longer guaranteed to be optimal. In Chap. 3 we consider a net present value problem in which the resource allocation is fixed, while the task start times can be chosen as a wait-and-see decision.

We close this section with an overview of the literature on temporal networks under uncertainty. Detailed reviews of specific topics will be provided in later chapters.

Although temporal networks have been analyzed for more than 50 years, see for example [Ful61, Kel61, MRCF59], the literature on temporal networks under uncertainty is surprisingly sparse. Until recently, most research on temporal networks under uncertainty assumed a fixed resource allocation and focused on the makespan of the early start schedule. Following the classification in [MÖ01], we can categorize the literature into methods that identify "critical" tasks or task paths [Elm00], simulation techniques to approximate the makespan distribution [AK89], approaches that bound the expected makespan [BM95, BNT02, MN79], and methods that bound the cumulative distribution function of the makespan [LMS01, MÖ01].

Optimization problems that maximize a network's net present value under uncertainty generally model the task start times as a wait-and-see decision, while the resource allocation is assumed to be fixed. The problem has been approximated by a two-stage recourse model in [Ben06], where an optimal delay policy is sought that prescribes how long each task should be delayed beyond its earliest start time. Under the assumption that the task durations are independent and exponentially distributed, the net present value maximization problem is formulated as a continuous-time Markov decision process in [BR97, TSS06]. Finally, approximate solutions for net present value maximization problems have been obtained with a number of heuristics, see [Bus95, OD97, Ö98, TFC98, WWS00]. For an overview of net present value maximization problems in temporal networks, see [HDD97].

Makespan minimization problems under uncertainty typically assume that a resource assignment is selected here-and-now, while the task start times are modeled as a wait-and-see decision. For non-renewable resources, the makespan minimization problem has been formulated as a two-stage recourse model in [Wol85] and as a robust optimization problem in [CGS07, CSS07, JLF07, LJF04]. Except for [CGS07], all of these contributions model the resource assignment as a here-and-now decision. A makespan minimization problem with renewable resources is studied in [MS00].

For reviews of different aspects of optimization problems in temporal networks under uncertainty, see [AK89, BKPH05, Elm05, HL04, HL05, JW00, LI08, MÖ1, Pin08, Sah04].

Chapter 3
Maximization of the Net Present Value

3.1 Introduction

This chapter studies a temporal network whose tasks give rise to cash flows. Positive cash flows denote cash inflows (e.g., received payments), whereas negative cash flows represent cash outflows (e.g., accrued costs). We present a method that maximizes the network's net present value (NPV), which is defined as the discounted sum of all arising cash flows. NPV maximization problems arise in project management, process scheduling and several other application areas. For example, in capital-intensive IT and construction projects, large amounts of money are invested over long periods of time, and the wise coordination of cash in- and outflows crucially affects the profitability of such projects. In this context, the NPV can be regarded as the "cash equivalent" of undertaking a project.

We consider temporal networks whose task durations and cash flows are described by a discrete set of alternative scenarios with associated occurrence probabilities. Since the cash flows can be positive or negative, the early start policy (1.2) does not yield an optimal solution, see Sect. 1.1. Similarly, the choice of scenario-independent task start times frequently leads to infeasible schedules or severe losses in revenues. Net present value maximization problems under uncertainty are most commonly modeled and solved via Markov decision processes, see Sect. 2.2.3. While this approach allows to determine truly adaptive schedules that react to the uncertainties revealed over time, the method quickly becomes computationally demanding for large networks. In this chapter, we present a method that aims to overcome this difficulty by determining an optimal (scenario-independent) *target processing time* (TPT) policy for the network tasks. In case task $i \in V$ could be started earlier than its TPT in the realized scenario, it will be postponed to its TPT. If, on the other hand, task i cannot be started at its TPT (because preceding tasks finish late), then it will be started as soon as possible thereafter. Following the terminology from Sect. 2.2.1, we solve a two-stage recourse problem in which the TPT policy is chosen here-and-now, whereas the factual task start times are modeled as a wait-and-see decision. The class of TPT policies is a strict subset of the class of

W. Wiesemann, *Optimization of Temporal Networks under Uncertainty*,
Advances in Computational Management Science 10,
DOI 10.1007/978-3-642-23427-9_3, © Springer-Verlag Berlin Heidelberg 2012

non-anticipative scheduling decisions. By restricting ourselves to this class, we can solve the NPV maximization problem for networks of nontrivial size. We allow for generalized precedence relations [EK90, EK92] but disregard resource restrictions. We discuss these assumptions in Sect. 3.3.

The remainder of this chapter is organized as follows. In the next section we summarize related literature. Section 3.3 formulates the problem that we aim to solve, while Sect. 3.4 describes the components of a branch-and-bound solution procedure. Section 3.5 presents and interprets the results of a numerical study. We conclude in Sect. 3.6.

3.2 Literature Review

Maximizing the NPV of a temporal network was first suggested in [Rus70].[1] The paper considers problem instances (G, ζ, d, β), where $G = (V, E)$ represents the structure of the temporal network, ζ_i the cash flow arising at the start time of task $i \in V$, d_i the duration of task i and $\beta = 1/(1+\alpha)$ the discount factor with internal rate of return $\alpha > 0$. An arc $(i, j) \in E$ represents a finish-start precedence between the tasks i and j, that is, task j must not be started before task i has been completed. The assumption that the cash flows are realized at the beginning of the tasks is not restrictive; we come back to this point in Sect. 3.3. All parameters are assumed to be deterministic, and there are no resource restrictions. The problem can be described as follows:

$$\begin{aligned}
\underset{y}{\text{maximize}} \quad & \sum_{i \in V} \zeta_i \beta^{y_i} \\
\text{subject to} \quad & y \in \mathbb{R}^n \\
& y_j \geq y_i + d_i \qquad \forall\, (i, j) \in E, \\
& y_1 = 0.
\end{aligned} \tag{3.1}$$

Note that this formulation is identical to the NPV maximization problem (1.3) discussed in Sect. 1.1. In problem (3.1), the components of the decision vector y represent the task start times. The objective function maximizes the sum of discounted cash flows. The constraints ensure satisfaction of the precedence relations and nonnegativity of the schedule. Without loss of generality, it is assumed that the first task is started at time zero. A deadline Δ can be imposed by adding the constraint $y_n \leq \Delta$. In [Rus70], problem (3.1) is solved through a sequence of linear programs whose objective functions are obtained by linearization around the current candidate solution. The duals of these approximations can be formulated as network flow problems, and the author proves local convergence of the overall procedure.

[1] We will use a consistent notation for all models reviewed in this section. Therefore, we may slightly modify some of the original formulations without changing their meaning.

It is shown in [Gri72] that the variable substitution $z_i = \beta^{y_i}$ converts problem (3.1) into an equivalent linear program:

$$\begin{aligned} \underset{z}{\text{maximize}} \quad & \sum_{i \in V} \zeta_i z_i \\ \text{subject to} \quad & z \in \mathbb{R}^n \\ & z_j \leq \beta^{d_i} z_i \qquad \forall (i,j) \in E, \\ & z_1 = 1, \\ & z_n \geq 0. \end{aligned}$$

A deadline Δ can be enforced by replacing the last constraint with $z_n \geq \beta^{\Delta}$. In [Gri72] this problem is solved with a network simplex variant.

In [NZ00], the network simplex algorithm from [Gri72] is extended to temporal networks $\Gamma = (G, \zeta, d, \beta)$ with generalized precedences. Here, $G = (V, E)$ represents the network structure, ζ_i the cash flow arising at the start time of task $i \in V$, d_{ij} the minimum time lag between the start times of tasks i and j ($d_{ij} < 0$ is allowed; hence, the precedences are called "generalized") and β the discount factor.[2] The authors solve problems of the following type:

$$\begin{aligned} \underset{y}{\text{maximize}} \quad & g(y) = \sum_{i \in V} \zeta_i \beta^{y_i} \\ \text{subject to} \quad & y \in \mathbb{R}^n \\ & y_j \geq y_i + d_{ij} \qquad \forall (i,j) \in E, \\ & y_1 = 0. \end{aligned} \tag{3.2}$$

It is shown in [DH02, SZ01] that the algorithm presented in [NZ00] performs favorably in practice. We will use this algorithm to solve subproblems that arise in the branch-and-bound procedure from Sect. 3.4.

In [EH90], an approximate solution procedure for the NPV maximization problem is proposed, while an exact method based on the steepest ascent principle is developed in [SZ01]. Comparisons of the various approaches can be found in [DH02, SZ01].

Over the last two decades, numerous publications have addressed extensions of the deterministic NPV maximization problem, the majority of which allow for different types of resource constraints. We do not provide more details on those efforts and refer the interested reader to the extensive surveys [DH02, HDD97]. Interestingly, the incorporation of uncertainty has attracted significantly less attention, just as temporal networks under uncertainty have in general been neglected for a long time.

---

[2]Generalized precedences are explained further in Sect. 3.3.

Assuming independent and exponentially distributed task durations, a stochastic version of problem (3.1) is considered in [BR97,TSS06]. Both contributions employ continuous-time Markov chains whose states assign labels "not yet started", "in progress" and "finished" to all tasks. Continuous-time Markov chains were first applied to temporal networks in [KA86]. The restriction to exponentially distributed durations can be relaxed at the cost of augmenting the state space. However, the number of states in the Markov chain grows exponentially with the network size, even when assuming exponentially distributed task durations. As a result, these methods are primarily applicable to small networks.

Several heuristics have been suggested for general task duration distributions. A suboptimal task delay policy is determined via simulation-based optimization in [Bus95]. The author points out that the problem is very challenging since the objective function is highly variable but flat near the (suspected) optimum. A simulated annealing heuristic for discretized task duration distributions is presented in [WWS00]. A general solution approach for stochastic temporal networks based on floating factor policies can be found in [TFC98]. The authors define the total float of task $i \in V$ as the difference between its latest (κ_i) and earliest (λ_i) start times given some deadline and average task durations. For a fixed float factor $\alpha \in [0, 1]$, task i should be started as early as possible but not before time $\lambda_i + \alpha(\kappa_i - \lambda_i)$. It is suggested to evaluate the impact of α via Monte Carlo simulation and to choose the value of α that minimizes a composite risk measure (e.g., the probability that the makespan or the overall costs exceed specified tolerances).

The method that comes closest to the one presented in this chapter is developed in [Ben06]. Again, the network structure is assumed to be given as $G = (V, E)$, where an arc $(i, j) \in E$ stipulates that task j cannot be started before task i has been finished. The author assumes that the cash flows are deterministic, whereas finitely many scenarios $s \in S$ with associated occurrence probabilities p_s specify the uncertain task durations d_i^s, $i \in V$. The goal is to maximize the expected net present value over all scenarios, which is done heuristically by determining a delay policy with the following two-stage procedure. In the first stage, the optimal "average" task start times are approximated through the solution of the following optimization problem:

$$\begin{aligned}
\underset{y}{\text{maximize}} \quad & \sum_{i \in V} \zeta_i \beta^{y_i} \\
\text{subject to} \quad & y \in \mathbb{R}^n \\
& y_j \geq \sum_{s \in S} p_s \max_{i \in V} \{y_i + d_i^s \, : \, (i, j) \in E\} \qquad \forall \, j \in V \setminus \{1\}, \\
& y_1 = 0.
\end{aligned}$$

In this model, cash flow ζ_i is assumed to arise at the start time y_i of task $i \in V$; we refer to Sect. 3.3 for a further discussion. The precedence constraints are imposed to hold in expectation. This model is not convex, and the author uses a local search procedure to obtain locally optimal start times. In the second stage, a task delay policy r is determined by setting $r_1 = y_1^*$ and

$$r_j = y_j^* - \sum_{s \in S} p_s \max_{i \in V} \{y_i^* + d_i^s \, : \, (i,j) \in E\} \qquad \forall \, j \in V \setminus \{1\},$$

where y^* denotes an optimal solution to the first-stage problem. The fixed-delay policy r prescribes to delay the start time of task $i \in V$ by r_i time units (compared to its earliest possible start time, which itself depends on the realized scenario). This approach is very attractive from a computational point of view, but it does not give any guarantees with respect to optimality. We will revisit this method in Sect. 3.5 when we compare its solutions with schedules obtained from TPT policies.

Finally, we mention the contributions [OD97,Ö98], which maximize a network's NPV subject to capital constraints and multiple task execution modes. Capital is treated as a randomly replenished resource which can be temporarily acquired at given costs. The authors present an online scheduling heuristic to solve this problem.

3.3 Problem Formulation

We study temporal networks in activity-on-node notation (see Sect. 1.1) with generalized precedence relations. This means that we allow for both minimum and maximum time lags between the start and completion times of the network tasks. A minimum time lag of length $\delta \geq 0$ between the start times of tasks i and j is modeled as a precedence relation $(i,j) \in E$ with positive value $d_{ij} = \delta$, whereas a similar maximum time lag of length $\delta \geq 0$ corresponds to a precedence relation $(j,i) \in E$ with negative value $d_{ji} = -\delta$. This allows us to represent both minimum and maximum time lags by inequalities of type $y_q \geq y_p + d_{pq}$, $(p,q) \in E$, where y_p and y_q represent the start times of tasks p and q, respectively. Since the completion time of a task equals its start time plus its duration, this approach immediately extends to time lags specified in terms of both start and completion times. A finish-start precedence between tasks i and j (i.e., "j cannot start before i has been completed") reduces to a minimum time lag of value 0 between the completion time of i and the start time of j. We refer to [EK90,EK92] for a detailed discussion of generalized precedence relations, together with convenient ways of specifying temporal networks in this format.

We consider problem instances $\Upsilon = (G, S, p, \zeta, d, \Delta, \beta)$, where $G = (V, E)$ represents the network structure and $S = \{1, \ldots, m\}$ the index set of discrete scenarios with occurrence probabilities p_s for $s \in S$. ζ_i^s denotes the cash flow arising at the start time of task $i \in V$ in scenario $s \in S$. The assumption that cash flows arise at the task start times is not restrictive: imagine, for example, that a cash flow z_i^s in scenario $s \in S$ arises when task $i \in V$ is *completed*. Assuming that the discount factor is β and that the duration of task i amounts to δ_i^s in scenario s, the end-of-task cash flow z_i^s is equivalent to a cash flow $\zeta_i^s = \beta^{\delta_i^s} z_i^s$ at the start time of task i in scenario s. The value of precedence relation $(i, j) \in E$ in scenario $s \in S$ is denoted by d_{ij}^s. Without loss of generality, we assume that for a given precedence $(i, j) \in E$, d_{ij}^s is of equal sign for all $s \in S$. We can then define the subset of positive-valued and negative-valued precedence relations by $E^+ = \{(i, j) \in E : d_{ij}^s \geq 0 \; \forall \, s \in S\}$ and $E^- = E \setminus E^+$, respectively. In Sect. 1.1 we stipulated that task 1 (n) constitutes the unique source (sink) of the network. In the light of generalized precedence relations, we now impose the same requirements for the subgraph $G = (V, E^+)$. In order to avoid unbounded problem instances, that is, instances in which it is beneficial to delay some network tasks indefinitely, we assume that there is a scenario-independent deadline Δ. Since we can choose Δ as large as we wish, this assumption does not restrict the generality of the approach. As before, β denotes the discount factor.

With the notation introduced above, the model we intend to solve can be formulated as follows:

$$\underset{r,y}{\text{maximize}} \quad f(r, y) = \sum_{s \in S} p_s \sum_{i \in V} \zeta_i^s \beta^{y_i^s} \tag{3.3a}$$

$$\text{subject to} \quad r \in \mathbb{R}^n, \quad y \in \mathbb{R}^{nm}$$

$$y_j^s = \max \left\{ \sup_{i \in V} \left\{ y_i^s + d_{ij}^s \; : \; (i, j) \in E^+ \right\}, r_j \right\} \qquad \forall \, j \in V, \, s \in S, \tag{3.3b}$$

$$y_j^s \geq y_i^s + d_{ij}^s \qquad \forall \, (i, j) \in E^-, \, s \in S, \tag{3.3c}$$

$$y_n^s \leq \Delta \qquad \forall \, s \in S, \tag{3.3d}$$

$$r_j \geq 0 \qquad \forall \, j \in V. \tag{3.3e}$$

For future use, we also define

$$\mathcal{Y}_\Upsilon = \{(r, y) \in \mathbb{R}^n \times \mathbb{R}^{nm} \; : \; (r, y) \text{ satisfies (3.3b)–(3.3e)}\}.$$

In model (3.3), r represents the desired TPT policy. This policy is chosen here-and-now, that is, before the realized scenario is known. The wait-and-see decision y_i^s denotes the factual start time of task $i \in V$ if scenario $s \in S$ is realized. Since the cash flows are realized at the task start times, the objective function represents the expected NPV over all scenarios. The constraint (3.3b) uniquely specifies the task start times y as a function of preceding start times and the TPT policy r. In particular, the start time of task j in scenario s only depends on the start times of preceding tasks i, $(i, j) \in E^+$, in this scenario, the respective minimum time lags and r_j. Note that the value of vector y is uniquely determined by r and d. Hence, y is not a decision vector in the ordinary sense, but it rather constitutes an ancillary variable vector that is required to evaluate the expected NPV over all scenarios. The constraint (3.3c) ensures satisfaction of the negative-valued precedence relations, while the constraint (3.3d) enforces the deadline in all scenarios. Note that the constraint (3.3d) cannot be replaced with maximum time lags between events

1 and n since y_1^s is allowed to be strictly positive (by choosing $r_1 > 0$). The constraint (3.3e) ensures nonnegativity of the solution. We remark that r can be chosen freely as long as the corresponding vector y satisfies all precedence relations in every scenario. Fixing r to the zero vector, for example, entails that all tasks are started as early as possible in every scenario. The relation enforced between r and y as described by the constraint (3.3b) is in accordance with our definition of TPT policies (see Sect. 3.1), and it constitutes a sufficient condition for non-anticipativity of the solution. However, it is not a necessary condition for non-anticipativity, and there might be feasible scheduling policies that result in better solutions.

Model (3.3) employs the expected value as decision criterion, which is in line with the majority of contributions for NPV maximization under uncertainty. Sometimes, however, cautious decision makers might take a more conservative stance and wish to avoid excessive losses in any particular scenario. Model (3.3) can account for individual risk preferences if the task cash flows ζ are replaced with associated utilities. In this case, the model maximizes the expected (discounted) overall utility [Fis70]. Note also that the described model allows both the cash flows and the task durations to depend on the realized scenario. This is desirable as longer task durations typically imply higher task costs, which themselves have a direct impact on the associated cash flows. In accordance with the existing body of literature, the model disregards resource constraints and assumes a constant discount factor β. Absence of resource restrictions constitutes a compromise that facilitates tractability of the resulting model. Apart from computational considerations, one could justify the absence of resource restrictions by the fact that NPV maximization models are typically employed in the early stages of the planning process to evaluate the profitability of an investment opportunity. At this stage, resource constraints may be of minor concern and can sometimes be dealt with by managerial intervention (e.g., in the context of project management by acquiring additional resources, shifting holidays or relying on overtime).

A possible variation of model (3.3) is to find optimal task delays in the spirit of [Ben06], see Sect. 3.2. Instead of target processing times, we would then seek for a scenario-independent task delay policy that specifies how much to defer task j beyond the expiry of minimum time lags $(i, j) \in E^+$. The resulting formulation is neither a special case nor a generalization of model (3.3): for a given problem instance, either model can lead to a superior expected NPV. In view of exact solution procedures, however, such a task delay formulation seems significantly more involved than model (3.3).[3] In Sect. 3.5.1, we compare TPT policies with task delay policies obtained from the two-stage procedure developed in [Ben06]. Finally, if the durations in model (3.3) do not depend on the realized scenario, then the optimal TPT policy can be determined by solving a deterministic NPV maximization problem with averaged cash flows $\zeta_i' = \sum_{s \in S} p_s \zeta_i^s$. This follows from the linearity of the expectation operator.

[3]The reason for this becomes clear when we discuss the nodal bounds of the branch-and-bound scheme in Sect. 3.4.1. While the bounds for model (3.3) are determined through linear programs, the bounding problems that arise in the task delay formulation constitute nonconvex problems.

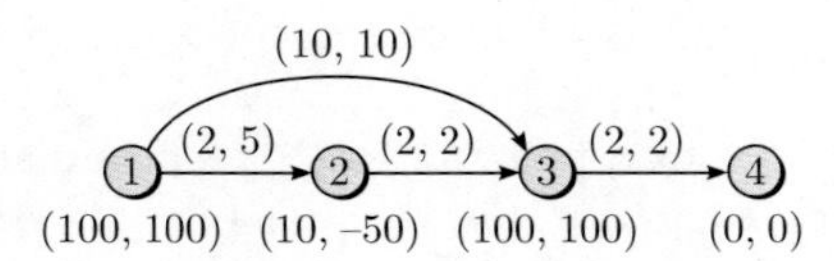

Fig. 3.1 Stochastic NPV maximization problem with two scenarios. The numbers attached to the arcs denote the values of the precedences, while the numbers attached to the nodes represent cash flows. In both cases, the first (second) number refers to the value in scenario 1 (2)

Let us examine the convexity properties of model (3.3). Leaving the objective function aside for the moment (we could potentially linearize it by using the variable substitution from [Gri72]), only the constraint (3.3b) requires investigation. Its right-hand sides are convex but generically not affine as they constitute maxima of affine functions. Thus, the constraint (3.3b) leads to a nonconvex set of feasible solutions. We cannot replace the equalities by greater or equal constraints, however, as otherwise non-anticipativity can be violated in the presence of network tasks with cash outflows. To illustrate this, let us assume that $\zeta_j^s < 0$ for $(j, s) \in V \times S$. Ceteris paribus, it would be beneficial to start task j in scenario s at the latest possible time consistent with all precedence relations, that is, at time $\inf_{k \in V} \{y_k^s - d_{jk}^s : (j, k) \in E\}$. As this time can exceed both $\max_{i \in V} \{y_i^s + d_{ij}^s : (i, j) \in E^+\}$ and r_j, such a decision would anticipate the realized scenario and as such violate causality.

The nonconvexity of problem Υ can also be illustrated by the temporal network in Fig. 3.1. In this network, node 4 represents a dummy task that signalizes the completion of all tasks. For two scenarios, $p = (0.5, 0.5)$, $\Delta = 20$ and any $\beta > 0$, the scenario-wise optimal (i.e., anticipative) solutions y^1 and y^2 are visualized on the left side of Fig. 3.2. We see that task 2 is started as early as possible in scenario 1 as it leads to a cash inflow. The same task starts as late as possible in scenario 2, however, since it leads to a cash outflow there. The right part of Fig. 3.2 shows the (non-anticipative) schedules stipulated by TPT vector $r = (0, 0, 0, 0)$. Set $\varphi(\lambda) = f((0, \lambda, 0, 0), y_\lambda)$, $\lambda \in [0, 16]$, where y_λ denotes the unique task start time vector that satisfies $((0, \lambda, 0, 0), y_\lambda) \in \mathcal{Y}_\Upsilon$ for a given λ. For a sufficiently large β, φ has zero slope for $\lambda \in [0, 2)$ (changing r_2 has no impact), a negative slope for $\lambda \in (2, 5)$ (the start time of task 2 is postponed in scenario 1), a positive slope for $\lambda \in (5, 8)$ (task 2 is postponed in both scenarios), and finally a negative slope for $\lambda \in (8, 16]$ (tasks 2–4 are postponed in both scenarios). Thus, the network's NPV is neither convex nor concave in r.

The solution approach that we present in this chapter requires the scenario set S to be of small cardinality, that is, it should not contain more than 20–30 elements. While this may be seen as a limitation of the method, we remark that stochastic NPV maximization problems are known to be challenging [HDD97, HL05]. As an alternative to the presented approach, one could try to employ a scenario-free uncertainty model. Popular scenario-free approaches to optimization problems in temporal networks are based on exponentially distributed task durations, see [BR97, TSS06], or employ a min–max objective, see Chap. 6 and [CGS07, CSS07]. However, both approaches lead to challenging optimization problems themselves, the former one

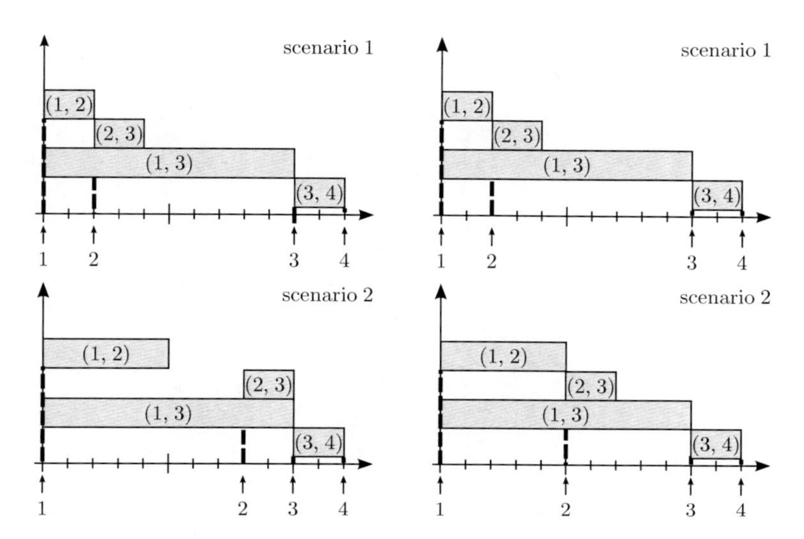

Fig. 3.2 Gantt charts for the scenario-wise optimal schedule (*left*) and the TPT schedule $r = (0, 0, 0, 0)$ (*right*). The horizontal axis displays the elapsed time, while the vertical axis lists the precedence relations. Arrows indicate the task start times

due to the curse of dimensionality in dynamic programming and the latter one due to the nonconvexity and two-stage nature of model (3.3). In the future, heuristic solution procedures may help to tackle instances of problem (3.3) with larger scenario sets. The (exact) solution procedure discussed in this chapter can then be used to assess the performance of such heuristics.

Suitable task duration and cash flow scenarios can be obtained from task-wise estimates or via scenario planning techniques. In the former case, alternative outcomes for the duration and the cash flow of a given task can be determined, for example, by employing three point estimates as in the classical PERT model [MRCF59]. The scenario set S then results from the cross product of all individual task outcomes. Clearly, this approach produces huge scenario sets: even if we assume that every task contributes only three different scenarios, we end up with a scenario set of cardinality $3^{|V|}$. However, scenario reduction techniques may be used to determine a small subset of scenarios that describes the aforementioned cross product as well as possible [HKR09, HR03]. Scenario planning techniques, on the other hand, ask the decision maker to identify the key drivers that affect the durations and cash flows of *all* (or many) tasks. In the context of project management, key drivers could be the weather, commodity prices and future exchange rates. One can then construct an initial set of scenarios by attaching probabilities to the various combinations of possible driver outcomes (e.g., via cross-impact analysis [GH68]). The number of scenarios can subsequently be reduced by clustering techniques. Scenario planning techniques have gained popularity in both theory [KY97, Sch01] and practice [Sch95].

We close this section with an example that illustrates formulation (3.3).

Example 3.3.1. Consider the temporal network in Fig. 3.1, and assume that $p = (0.5, 0.5)$, the deadline is $\Delta = 20$, and the discount factor is $\beta = 0.95$. In this case, model (3.3) becomes

$$\begin{aligned}
\underset{r,y}{\text{maximize}} \quad & 1/2\left(100 \cdot 0.95^{y_1^1} + 10 \cdot 0.95^{y_2^1} + 100 \cdot 0.95^{y_3^1}\right) \\
& + 1/2\left(100 \cdot 0.95^{y_1^2} - 50 \cdot 0.95^{y_2^2} + 100 \cdot 0.95^{y_3^2}\right) \\
\text{subject to} \quad & r \in \mathbb{R}^4, \quad y \in \mathbb{R}^8 \\
& y_1^1 = r_1, \quad y_2^1 = \max\left\{y_1^1 + 2,\ r_2\right\}, \\
& y_3^1 = \max\left\{y_1^1 + 10,\ y_2^1 + 2,\ r_3\right\}, \quad y_4^1 = \max\left\{y_3^1 + 2,\ r_4\right\}, \\
& y_1^2 = r_1, \quad y_2^2 = \max\left\{y_1^2 + 5,\ r_2\right\}, \\
& y_3^2 = \max\left\{y_1^2 + 10,\ y_2^2 + 2,\ r_3\right\}, \quad y_4^2 = \max\left\{y_3^2 + 2,\ r_4\right\}, \\
& y_4^1 \leq 20, \quad y_4^2 \leq 20, \\
& r_1, r_2, r_3, r_4 \geq 0.
\end{aligned}$$

As we will show later in Example 3.4.1, this model is optimized by $\widehat{r} = (0, 8, 0, 0)$ and $(\widehat{y}_1^s, \widehat{y}_2^s, \widehat{y}_3^s, \widehat{y}_4^s) = (0, 8, 10, 12)$, $s \in \{1, 2\}$. Thus, the optimal TPT policy assigns a target processing time of 8 to task 2, while all other tasks should be started as early as possible. The factual task start times for this policy are 0, 8, 10 and 12 for tasks 1, 2, 3 and 4, respectively, and they do not depend on the realized scenario. The optimal objective value is $f(\widehat{r}, \widehat{y}) = 116.96$.

Assume now that in either scenario, task 3 must start at most 7 time units after task 1 has been started. We achieve this by adding a precedence $(3, 1)$ to E with duration $(d_{31}^1, d_{31}^2) = (-7, -7)$. In this case, we would add the constraints

$$y_1^1 \geq y_3^1 - 7, \quad y_1^2 \geq y_3^2 - 7$$

to the optimization model. The TPT policy $\widehat{r}$ is no longer feasible under this additional constraint.

3.4 Solution Procedure

In the following, we present a branch-and-bound procedure for the solution of model (3.3). Branch-and-bound algorithms solve optimization problems by implicitly enumerating the set of feasible solutions in a branch-and-bound tree $\mathcal{T}$. Every node of $\mathcal{T}$ represents a subset of the feasible solutions. The tree construction starts at the root node, which represents the entire set of feasible solutions. Branch-and-bound algorithms iteratively select tree nodes $\tau \in \mathcal{T}$ for branching. When a node τ is branched, its set of feasible solutions, $\mathcal{Y}_\tau$, is split into several subsets whose union coincides with $\mathcal{Y}_\tau$. Every subset thus generated represents a "child" node of τ in $\mathcal{T}$. In principle, nodes of $\mathcal{T}$ can be split until their associated solution sets reduce (or converge) to singletons. In order to avoid such a complete enumeration, one calculates bounds on the optimal objective value achievable at each tree node. A node may then be fathomed as soon as it is guaranteed that it does not contain any better solution than the best one currently known. The crucial components of branch-and-bound procedures are the employed bounds, the branching scheme and the node selection rule, that is, a recipe that specifies which node of $\mathcal{T}$ to split next.

In the branch-and-bound algorithm presented in this chapter we determine an upper bound on the optimal objective value achievable at tree node τ by maximizing f, the objective function of (3.3), over a relaxation of $\mathcal{Y}_\tau$ that neglects non-anticipativity. This is done by replacing the equalities in (3.3b) by greater or equal relations. As long as the optimal solution (r, y) to such a relaxation contains a variable y_j^s that satisfies the strict inequality

$$y_j^s > \max\left\{\sup_{i\in V}\left\{y_i^s + d_{ij}^s \,:\, (i,j)\in E^+\right\}, r_j\right\},$$

non-anticipativity is violated (see Sects. 2.2.1 and 3.3). In this case, the branching scheme fixes y_j^s to either r_j or to $y_i^s + d_{ij}^s$ for a task $i \in V$ that satisfies $(i,j) \in E^+$, and every such fixation leads to a child node of τ.

We now formalize this idea. The relaxed feasible set $\mathcal{Z}_\Upsilon$ is defined through

$$\left.\begin{array}{ll} y_j^s \geq y_i^s + d_{ij}^s & \forall\,(i,j)\in E,\ s\in S,\\ y_j^s \geq r_j & \forall\, j\in V,\ s\in S,\\ y_n^s \leq \Delta & \forall\, s\in S,\\ r_j \geq 0 & \forall\, j\in V \end{array}\right\} \quad \Leftrightarrow \quad (r,y)\in \mathcal{Z}_\Upsilon. \tag{3.4}$$

Note that $\mathcal{Z}_\Upsilon$ constitutes a convex relaxation of $\mathcal{Y}_\Upsilon$ as defined in (3.3). The requirement that y_j^s has to equal $y_i^s + d_{ij}^s$ for some $i \in V$ with $(i,j) \in E^+$ or r_j will be enforced by restricting (r, y) to one of the hyperplanes

$$\mathcal{Z}_{ij}^s = \begin{cases} \left\{(r,y) \,:\, y_j^s = y_i^s + d_{ij}^s\right\} & \text{for } (i,j)\in E^+,\ s\in S,\\ \left\{(r,y) \,:\, y_j^s = r_j\right\} & \text{for } i = j\in V,\ s\in S,\\ \emptyset & \text{otherwise.} \end{cases} \tag{3.5}$$

We identify a tree node $\tau \in \mathcal{T}$ with the hyperplane restrictions it enforces, that is, $\tau \subseteq V^2 \times S$. For a given node τ, we define the set of feasible solutions, $\mathcal{Y}_\tau$, as well as its relaxation, $\mathcal{Z}_\tau$, as

$$\mathcal{Y}_\tau = \mathcal{Y}_\Upsilon \cap \bigcap_{(i,j,s)\in\tau} \mathcal{Z}_{ij}^s \quad \text{and} \quad \mathcal{Z}_\tau = \mathcal{Z}_\Upsilon \cap \bigcap_{(i,j,s)\in\tau} \mathcal{Z}_{ij}^s. \tag{3.6}$$

$\mathcal{Y}_\Upsilon \subseteq \mathcal{Z}_\Upsilon$ implies that $\mathcal{Y}_\tau \subseteq \mathcal{Z}_\tau$, and hence we obtain an upper bound on the achievable objective value at node τ by maximizing f over $\mathcal{Z}_\tau$ instead of $\mathcal{Y}_\tau$.

Given a TPT policy $r \in \mathbb{R}_+^n$, we define the *induced schedule* $y(r) \in \mathbb{R}^{nm}$ recursively as follows.

$$y_j^s(r) = \max\left\{\sup_{i\in V}\left\{y_i^s(r) + d_{ij}^s \ : \ (i,j) \in E^+\right\}, r_j\right\}. \tag{3.7}$$

Remember that the precedence relations in E^+ are nonnegative valued in all scenarios $s \in S$. If there is a cycle in E^+, then all arcs in the cycle must be associated with zero-valued precedences, for otherwise the network structure is inconsistent (a task cannot be completed before it is started). Thus, the induced schedule is well defined. Due to the relation between (3.7) and (3.3b), $y_j^s(r)$ equals the factual start time of task j if the TPT policy r is implemented and scenario s is realized. By construction, $y(r)$ satisfies all minimum time lags in every scenario. If $y(r)$ also satisfies all maximum time lags in every scenario, that is, if $(r, y(r)) \in \mathcal{Y}_\Upsilon$, then we call r a *feasible policy*.

With this notation, the solution approach can be described as Algorithm 1.

Instead of storing the branch-and-bound tree $\mathcal{T}$ explicitly, the algorithm keeps a list $\mathcal{L}$ of nodes that have been constructed by Steps 1 and 4 but not yet selected by Step 2. The relation between $\mathcal{T}$ and $\mathcal{L}$ is that $\tau \in \mathcal{T}$ if and only if $\tau \in \mathcal{L}$ at some point during the execution of the algorithm.

In *Step 1*, $\mathcal{L}$ only contains the root node $\tau_0 = \emptyset \subseteq V^2 \times S$. Hence, all non-anticipativity constraints in (3.3b) are relaxed in the beginning. We use r^* and f^* to record the best TPT policy found so far and its expected NPV. We assign an expected NPV of $-M$ to infeasible policies.

In *Step 2*, we first check whether we can terminate. This is the case if no more nodes are available or eligible for further processing. Nodes are *available* for further processing if $\mathcal{L}$ is nonempty. Note that after Step 1, $\mathcal{L}$ contains the root node $\tau_0 = \emptyset \subseteq V^2 \times S$ and thus is *not* empty. A node is *eligible* for further processing if its upper bound exceeds f^*, that is, if it can contain a better TPT policy than r^*. If $\mathcal{L}$ contains eligible nodes, then we select a node that attains the maximal upper bound. In the following, we refer to this node as τ.

In *Step 3*, we calculate an upper bound on the maximal value of f over $\mathcal{Y}_\tau$. This bound is determined by the maximal value of f over the relaxed constraint set $\mathcal{Z}_\tau$ as defined in (3.6), and the corresponding optimal solution is denoted by $(\widehat{r}, \widehat{y})$. If $\widehat{r}$ constitutes a feasible TPT policy, then we also obtain a lower bound $f(\widehat{r}, y(\widehat{r}))$ on the (globally) best TPT policy. We use this lower bound to improve (r^*, f^*) if possible. Note that $\widehat{y} \neq y(\widehat{r})$ in general: $\widehat{y}$ contains the optimal task start times when neglecting some of the non-anticipativity constraints. As such, $\widehat{y}$ typically

Algorithm 1 Branch-and-bound scheme for model (3.3).

1. **Initialization.** Set $\mathcal{L} = \{\tau_0\}$ with $\tau_0 = \emptyset$, $r^* = 0$, $f^* = f(r^*, y(r^*))$ if r^* is feasible and $f^* = -M$ otherwise.[4]
2. **Node Selection.** If $\mathcal{L} = \emptyset$ or
$$\max_{\tau \in \mathcal{L}} \sup_{(r,y) \in \mathcal{Z}_\tau} f(r, y) \leq f^*,$$
then go to Step 5. Otherwise, select a node $\tau \in \mathcal{L}$ with
$$\tau \in \arg\max_{\tau \in \mathcal{L}} \sup_{(r,y) \in \mathcal{Z}_\tau} f(r, y)$$
and set $\mathcal{L} = \mathcal{L} \setminus \{\tau\}$.
3. **Bounding.** For the node τ selected in Step 2, let
$$(\widehat{r}, \widehat{y}) = \arg\max \{f(r, y) \,:\, (r, y) \in \mathcal{Z}_\tau\}$$
be a solution to the upper bound problem at τ. If $\widehat{r}$ is feasible and
$$f(\widehat{r}, y(\widehat{r})) > f^*,$$
where $y(\widehat{r})$ is defined in (3.7), then set $r^* = \widehat{r}$ and $f^* = f(\widehat{r}, y(\widehat{r}))$, that is, a new incumbent TPT policy has been found.[5]
4. **Branching.** Let
$$V_\tau = \left\{(j, s) \in V \times S \,:\, \widehat{y}_j^s > \max\left\{\sup_{i \in V}\left\{\widehat{y}_i^s + d_{ij}^s \,:\, (i, j) \in E^+\right\}, \widehat{r}_j\right\}\right\}$$
be the set of task-scenario pairs violating non-anticipativity in $(\widehat{r}, \widehat{y})$. If $V_\tau \neq \emptyset$, then select $(j, s) \in V_\tau$ according to some branching scheme and set
$$\mathcal{L} = \mathcal{L} \cup \Big(\bigcup_{\substack{i \in V: \\ (i,j) \in E^+}} \{\tau \cup \{(i, j, s)\}\} \Big) \cup \{\tau \cup \{(j, j, s)\}\}.$$
Go to Step 2 (next iteration).
5. **Termination.** If $f^* \neq -M$, then r^* represents the optimal TPT policy. Otherwise, the problem is infeasible.

violates non-anticipativity, that is, the set V_τ defined in Step 4 is usually nonempty. The vector $y(\widehat{r})$, on the other hand, represents the task start times that result from implementing the TPT policy $\widehat{r}$ (see Sect. 3.1). Although $y(\widehat{r})$ is non-anticipative by construction, it may violate some negative-valued precedences $(i, j) \in E^-$. In model (3.3), the constraint (3.3b) ensures that feasible solutions $(r, y) \in \mathcal{Y}_\Upsilon$ satisfy $y = y(r)$. In the branch-and-bound algorithm, coincidence of $\widehat{y}$ and $y(\widehat{r})$ is established gradually by adding hyperplane restrictions $\mathcal{Z}_{ij}^s$.

If the upper bound solution $(\widehat{r}, \widehat{y})$ violates non-anticipativity, then we select an anticipating task-scenario pair $(j, s) \in V \times S$ in *Step 4* according to some branching scheme. Among several branching schemes, including rules based on task start times, numbers of incoming positive-valued precedences and gaps between task start times and their incoming positive-valued precedences, the following strategy turns out to perform best: for every anticipative task-scenario pair $(j, s) \in V \times S$, determine the minimum decrease in objective value caused by shifting j to the expiration time of any of its incoming positive-valued precedences or to r_j in scenario s:

$$\eta(j,s) = p_s \left|\zeta_j^s\right| \min\left\{ \inf_{i \in V} \left\{ \beta^{\widehat{y_i^s + d_{ij}^s}} - \beta^{\widehat{y_j^s}} \;:\; (i,j) \in E^+ \right\}, \beta^{\widehat{r_j}} - \beta^{\widehat{y_j^s}} \right\}.$$

For $j \in V$ and $s \in S$, $\eta(j, s)$ approximates the minimum additional expected costs of ensuring non-anticipativity for the task-scenario pair (j, s). We select the anticipative task-scenario pair (j, s) with maximal $\eta(j, s)$. The hope is that this greedy selection rule leads to a fast decrease of the nodal upper bounds. Having selected a pair $(j, s) \in V \times S$, we create one child node for every possible fixation of y_j^s to one of its predecessors $i \in V$, $(i, j) \in E^+$. We also create a child node that fixes y_j^s to r_j. These new child nodes $\tau \cup \{(i, j, s)\}$ and $\tau \cup \{(j, j, s)\}$ are appended to $\mathcal{L}$, and then we go back to Step 2.

After finitely many iterations, $\mathcal{L}$ does not contain any further available or eligible nodes in Step 2. At this point, the algorithm enters *Step 5* and delivers either an optimal TPT policy or establishes the infeasibility of (3.3).

The correctness of the branch-and-bound algorithm is proved in two steps. First, we show that the algorithm always terminates after a finite number of iterations. Afterwards, we show that if the algorithm terminates, then it provides the correct result. The proofs of these assertions require some additional notation. We already described the correspondence between the node list $\mathcal{L}$ and the implicitly generated branch-and-bound tree $\mathcal{T}$. For any node $\tau \in \mathcal{T}$ we denote the set of direct descendants by $D_{\mathcal{T}}(\tau)$. Thus, we have that $\tau' \in D_{\mathcal{T}}(\tau)$ if and only if τ' is added to $\mathcal{L}$ as a result of branching τ in Step 4 of the algorithm. We denote the set of all (transitive) descendants of τ in $\mathcal{T}$ by $D_{\mathcal{T}}^*(\tau)$, that is, $D_{\mathcal{T}}^*(\tau)$ contains all direct descendants of τ, all direct descendant of τ's direct descendants, etc. Similarly, $A_{\mathcal{T}}(\tau)$ denotes the set of direct ancestors of τ (a singleton). We have $\tau \in A_{\mathcal{T}}(\tau')$ if and only if $\tau' \in D_{\mathcal{T}}(\tau)$. Finally, $A_{\mathcal{T}}^*(\tau)$ refers to all (transitive) ancestors of τ in $\mathcal{T}$.

We now prove finite termination and completeness of the algorithm.

Theorem 3.4.1 (Termination). *For any given problem instance, Algorithm 1 terminates after finitely many iterations.*

Proof. We show that the generated branch-and-bound tree $\mathcal{T}$ is finite. By the correspondence between $\mathcal{L}$ and $\mathcal{T}$, the claim then follows immediately. For every $\tau \in \mathcal{T}$, $|D_{\mathcal{T}}(\tau)|$ is bounded by $|V|$, because an anticipating task-scenario pair $(j, s) \in V \times S$ can only be fixed to either one of its preceding tasks $i \in V$ with $(i, j) \in E^+$ or to r_j. If $(j, s) \in V_\tau$ is branched upon, then $(j, s) \notin V_{\tau'}$ for any transitive descendant $\tau' \in D_{\mathcal{T}}^*(\tau)$, because either $(i, j, s) \in \tau'$ for $i \in V$ with $(i, j) \in E^+$ or $(j, j, s) \in \tau'$. As a result, no node $\tau \in \mathcal{T}$ can possess more than nm fixations. Hence, both the number of levels in $\mathcal{T}$ and the fan-out within each level are bounded, which proves finiteness of $\mathcal{T}$. □

Theorem 3.4.2 (Completeness). *Algorithm 1 returns $f^* = -M$ if the problem is infeasible and a TPT policy r^* with*

$$(r^*, y(r^*)) \in \arg\max \{f(r, y) : (r, y) \in \mathcal{Y}_\Upsilon\}$$

otherwise.

Proof. We first show that the algorithm correctly identifies infeasible instances. The algorithm classifies a problem as infeasible if $f^* = -M$ after termination. Note that f^* can only change in Step 3. For this to happen, however, r^* needs to be feasible, implying that a feasible solution in $\mathcal{Y}_\tau$ (and, a fortiori, in $\mathcal{Y}_\Upsilon$) has been found. Thus, for any infeasible instance, the algorithm returns $f^* = -M$, that is, it correctly recognizes the problem's infeasibility.

If instance Υ is feasible, then it has an optimal solution: the existence of a finite deadline, together with the assumption of a unique source in the subgraph (V, E^+), ensures that $\mathcal{Y}_\Upsilon$ is compact. The continuous function (3.3a) thus attains its maximum over $\mathcal{Y}_\Upsilon$ due to the Weierstrass maximum theorem.

We now prove that if the problem is feasible, then the algorithm finds an optimal solution. To show this, let r^{opt} be an optimal TPT policy. We examine the branch-and-bound tree $\mathcal{T}$ generated by the procedure. Let $\mathcal{T}' = \{\tau \in \mathcal{T} : (r^{\text{opt}}, y(r^{\text{opt}})) \in \mathcal{Y}_\tau\}$, that is, $\mathcal{T}'$ consists of all the tree nodes of $\mathcal{T}$ that contain the optimal solution $(r^{\text{opt}}, y(r^{\text{opt}}))$. Note that $\mathcal{T}' \neq \emptyset$ since it contains at least the root node of $\mathcal{T}$. We now remove from $\mathcal{T}'$ all nodes that have descendants in $\mathcal{T}'$, that is, $\mathcal{T}'' = \mathcal{T}' \setminus \bigcup_{\tau \in \mathcal{T}'} A^*_{\mathcal{T}'}(\tau)$. By construction, $\mathcal{T}'' \neq \emptyset$ holds as well. Let us fix an arbitrary $\tau \in \mathcal{T}''$. During the execution of the algorithm, τ has either been selected in Step 2 or not. If it has never been selected, then the inequality $f(r^{\text{opt}}, y(r^{\text{opt}})) \leq \max \{f(r, y) : (r, y) \in \mathcal{Z}_\tau\} \leq f^*$ must hold at the end of the algorithm, implying that a TPT policy at least as good as r^{opt} has been found. If τ has been selected in Step 2 at some point, then we know by definition of $\mathcal{T}''$ that τ has not been branched. This is only possible if $V_\tau = \emptyset$, that is, if $(\widehat{r}, \widehat{y}) \in \arg\max \{f(r, y) : (r, y) \in \mathcal{Z}_\tau\}$ is non-anticipative, where $(\widehat{r}, \widehat{y})$ denotes the upper bound for τ determined in Step 3. In that case, however, r^* has been updated to $\widehat{r}$ in Step 3 (if necessary) and thus, a TPT policy at least as good as r^{opt} has been identified by the method. □

The algorithm description does not contain any dominance rules. In fact, the dominance rules that prevail in the literature (see [BDM+99, DH02]) are based on partial schedules. The tree nodes of $\mathcal{T}$, on the other hand, represent sets of complete schedules which may not yet be feasible due to their anticipativity. As a result, classical dominance rules such as the "superset–subset" rule [DH02] are not (directly) applicable. Whether other dominance rules can be used beneficially to enhance the algorithm remains an area for further research.

Step 3 of the branch-and-bound procedure requires the efficient solution of

$$\max \{f(r, y) : (r, y) \in \mathcal{Z}_\tau\} \qquad (\Upsilon(\tau))$$

for nodes $\tau \in \mathcal{T}$, where $\mathcal{Z}_\tau$ results from the intersection of $\mathcal{Z}_\Upsilon$ with the hyperplanes indexed by τ, see (3.4)–(3.6). In the following, we refer to this problem as $\Upsilon(\tau)$. $\Upsilon(\tau)$ is equivalent to the following optimization problem:

$$\begin{aligned}
\underset{r,y}{\text{maximize}} \quad & \sum_{s \in S} p_s \sum_{i \in V} \zeta_i^s \beta^{y_i^s} \\
\text{subject to} \quad & r \in \mathbb{R}^n, \quad y \in \mathbb{R}^{nm} \\
& y_j^s \begin{cases} = y_i^s + d_{ij}^s & \text{if } (i,j,s) \in \tau \\ \geq y_i^s + d_{ij}^s & \text{otherwise} \end{cases} \qquad \forall\, (i,j) \in E,\ s \in S, \\
& y_j^s \begin{cases} = r_j & \text{if } (j,j,s) \in \tau \\ \geq r_j & \text{otherwise} \end{cases} \qquad \forall\, j \in V,\ s \in S, \\
& y_n^s \leq \Delta \qquad \forall\, s \in S, \\
& r_j \geq 0 \qquad \forall\, j \in V.
\end{aligned}$$

In analogy to the deterministic model (3.1), we could employ the substitutions $t_j = \beta^{r_j}$, $j \in V$, and $z_j^s = \beta^{y_j^s}$, $j \in V$ and $s \in S$, to transform this problem into an equivalent linear program. As we will show in Sect. 3.4.1, however, $\Upsilon(\tau)$ can also be reformulated as a deterministic NPV maximization problem. The latter approach improves the performance of the branch-and-bound procedure since the specialized algorithms reviewed in Sect. 3.2 outperform linear programming solvers by several orders of magnitude, see [SZ01]. In Sect. 3.4.2 we discuss how to exploit information from the father node in the branch-and-bound tree when solving $\Upsilon(\tau)$. This allows us to further speed up the calculation of nodal upper bounds as the algorithms reviewed in Sect. 3.2 require significantly fewer iterations when warm-started from near-optimal solutions.

We close this section with an illustration of the branch-and-bound procedure.

Example 3.4.1. Consider again the temporal network in Fig. 3.1 with scenario probabilities $p = (0.5, 0.5)$, deadline $\Delta = 20$ and discount factor $\beta = 0.95$. For this problem instance, the branch-and-bound algorithm proceeds as follows.

We start with *Step 1*, where we set $\mathcal{L} = \{\tau_0\}$ with $\tau_0 = \emptyset$ and $r^* = 0$. The induced schedule $y(r^*)$ is the early start schedule visualized in Fig. 3.2, right. Since there are no maximum time lags, this schedule is feasible and leads to an objective value of

$$\begin{aligned}
&1/2\left(100 \cdot 0.95^0 + 10 \cdot 0.95^2 + 100 \cdot 0.95^{10}\right) \\
&+1/2\left(100 \cdot 0.95^0 - 50 \cdot 0.95^5 + 100 \cdot 0.95^{10}\right) \approx 115.40.
\end{aligned}$$

Hence, we set $f^* = 115.40$.

The node $\tau = \tau_0 \in \mathcal{L}$ does not enforce any fixations yet. Hence, the problem $\sup_{(r,y)\in\mathcal{Z}_\tau} f(r,y)$ in *Step 2* is maximized by $\widehat{r} = 0$ and the anticipative task start time vector $\widehat{y}$ visualized in Fig. 3.2, left. The objective value is

$$1/2\left(100 \cdot 0.95^0 + 10 \cdot 0.95^2 + 100 \cdot 0.95^{10}\right)$$
$$+1/2\left(100 \cdot 0.95^0 - 50 \cdot 0.95^8 + 100 \cdot 0.95^{10}\right) \approx 118.16.$$

Since this value exceeds $f^* = 115.40$, we remove node τ_0 from $\mathcal{L}$ and continue.

In *Step 3* we check whether the objective value of the induced schedule $y(\widehat{r})$ exceeds the objective value of the induced schedule $y(r^*)$. Since $\widehat{r}$ and r^* are identical, this is not the case.

In *Step 4* we identify the set of anticipative task-scenario pairs for $\tau = \tau_0$ as $V_\tau = \{(2,2)\}$. We create two new nodes $\tau_1 = \{(1,2,2)\}$ ("start task 2 in scenario 2 precisely d_{12}^2 time units after task 1 has been started") and $\tau_2 = \{(2,2,2)\}$ ("start task 2 in scenario 2 at time r_2") and add them to $\mathcal{L}$.

We are back in *Step 2* with $\mathcal{L} = \{\tau_1, \tau_2\}$. For node $\tau = \tau_1$, the problem $\sup_{(r,y)\in\mathcal{Z}_\tau} f(r,y)$ is maximized by $\widehat{r} = 0$ and the early start schedule $\widehat{y}$ shown in Fig. 3.2, right. We have already seen that the associated objective value is $115.40 \leq f^* = 115.40$. For node $\tau = \tau_2$, the problem $\sup_{(r,y)\in\mathcal{Z}_\tau} f(r,y)$ is maximized by $\widehat{r} = (0,8,0,0)$ and the scenario-independent task start times $(\widehat{y}_1^s, \widehat{y}_2^s, \widehat{y}_3^s, \widehat{y}_4^s) = (0,8,10,12)$, $s \in \{1,2\}$. The objective value of this solution is

$$1/2\left(100 \cdot 0.95^0 + 10 \cdot 0.95^8 + 100 \cdot 0.95^{10}\right)$$
$$+1/2\left(100 \cdot 0.95^0 - 50 \cdot 0.95^8 + 100 \cdot 0.95^{10}\right) \approx 116.96.$$

Since $116.96 > f^* = 115.40$, we remove τ_2 from $\mathcal{L}$ and continue.

For node $\tau = \tau_2$, we obtain $y(\widehat{r}) = \widehat{y}$. Hence, $f(\widehat{r}, y(\widehat{r})) = 116.96$, and we update f^* and r^* in *Step 3* to $f^* = 116.96$ and $r^* = (0,8,0,0)$.

Since the solution $(\widehat{r}, \widehat{y})$ is non-anticipative, we obtain $V_\tau = \emptyset$ in *Step 4*. We do not branch τ_2.

Back in *Step 2*, the list $\mathcal{L}$ only contains node τ_1. Since the objective value of $\sup_{(r,y)\in\mathcal{Z}_\tau} f(r,y)$ for $\tau = \tau_1$ is 115.40 and $115.40 \leq f^* = 116.96$, no eligible nodes are left for branching.

We therefore go to *Step 5* and return the optimal TPT policy $r^* = (0,8,0,0)$. Figure 3.3 visualizes the branch-and-bound tree that we generated during the execution of the algorithm. Node τ_1 has been fathomed because its upper bound does not exceed the objective value of the incumbent solution, whereas τ_2 has been fathomed because its upper bound is attained by a feasible (i.e., non-anticipative) solution.

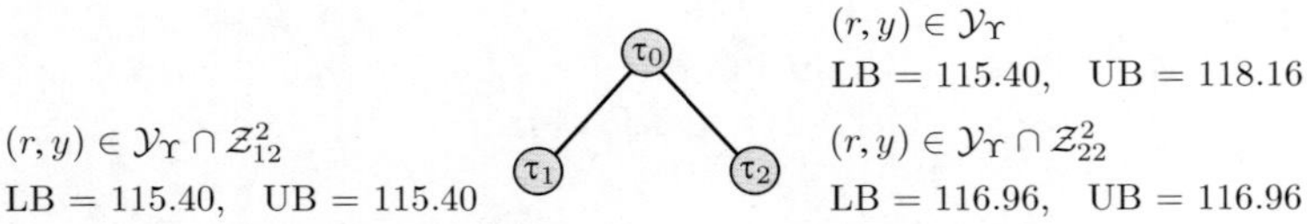

$(r, y) \in \mathcal{Y}_\Upsilon$
LB = 115.40, UB = 118.16

τ_0

τ_1 τ_2

$(r, y) \in \mathcal{Y}_\Upsilon \cap \mathcal{Z}^2_{12}$
LB = 115.40, UB = 115.40

$(r, y) \in \mathcal{Y}_\Upsilon \cap \mathcal{Z}^2_{22}$
LB = 116.96, UB = 116.96

Fig. 3.3 Branch-and-bound tree generated in Example 3.4.1. The nodal upper and lower bounds are denoted by "UB" and "LB", respectively

3.4.1 Efficient Nodal Bounds

Given a stochastic NPV maximization instance $\Upsilon = (G, S, p, \zeta, d, \Delta, \beta)$ and a tree node $\tau \in \mathcal{T}$, we define an instance $\Gamma(\tau) = (\widetilde{G}, \widetilde{\zeta}, \widetilde{d}, \beta)$ of the deterministic NPV maximization problem (3.2) as follows. $\widetilde{G} = (\widetilde{V}, \widetilde{E})$ represents a network whose node set $\widetilde{V} = \{0, \ldots, \widetilde{n}\}$ consists of three categories. The first category encompasses the artificial start node 0, which provides a unique source for the network. The second category consists of TPT nodes $i = 1, \ldots, n$, which correspond to the target processing times r_i, $i \in V$. The last category encompasses task-scenario nodes $p = sn + i$ for $i \in V$ and $s \in S$, which represent the task start times in the different scenarios. For a network with two scenarios, for example, nodes $n + 1, \ldots, 2n$ describe the start times of tasks $1, \ldots, n$ in scenario 1, respectively, and nodes $2n + 1, \ldots, 3n = \widetilde{n}$ describe the corresponding start times in scenario 2. We assign a cash flow of magnitude $p_s \zeta_i^s$ to task-scenario node $p = sn + i$, $i \in V$ and $s \in S$, while the other nodes in $\widetilde{V}$ do not give rise to cash flows. Thus, the node-related data of $\Gamma(\tau)$ can be summarized as follows:

$$\widetilde{V} = \{0, \ldots, \widetilde{n}\} \quad \text{with} \quad \widetilde{n} = n(m + 1), \tag{3.8a}$$

$$\widetilde{\zeta}_p = \begin{cases} p_s \zeta_i^s & \text{if } p = sn + i,\ i \in V \text{ and } s \in S, \\ 0 & \text{otherwise.} \end{cases} \tag{3.8b}$$

We now construct the precedences $\widetilde{E}$ of $\Gamma(\tau)$. We establish zero-valued precedences between the artificial start node 0 and all TPT nodes $i = 1, \ldots, n$ to ensure nonnegativity of the solution. Next, we add zero-valued precedences between the TPT nodes $i = 1, \ldots, n$ and the corresponding task-scenario nodes $p = sn + i$, $s \in S$. This guarantees that tasks are not started before their TPTs. For every precedence $(i, j) \in E$, we add to $\Gamma(\tau)$ a precedence of value d_{ij}^s between $p = sn + i$ and $q = sn + j$ for every scenario $s \in S$. This ensures that $\Gamma(\tau)$ obeys the original precedence relations of Υ in all scenarios. Fixations $(i, j, s) \in V^2 \times S$ are modeled as tight maximum time lags between the respective nodes in $\widetilde{V}$. Finally, satisfaction of the deadline Δ is ensured by adding maximum time lags of duration Δ between 0 and $sn + n$ for all scenarios $s \in S$.

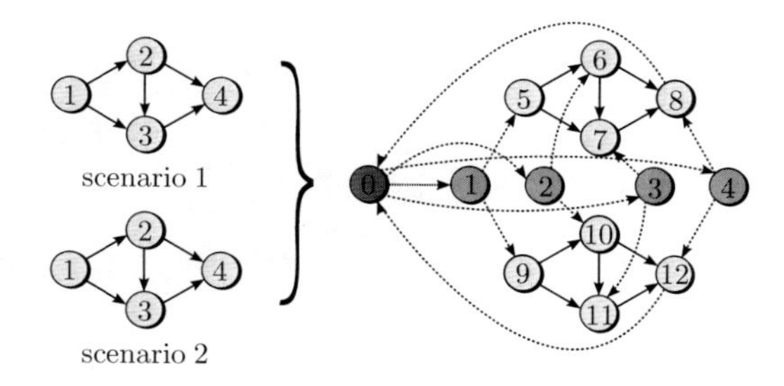

Fig. 3.4 For the stochastic NPV maximization instance in the *left chart* (only the network structure is shown), the deterministic NPV maximization problem $\Gamma(\tau_0)$ with $\tau_0 = \emptyset$ is visualized on the *right side*

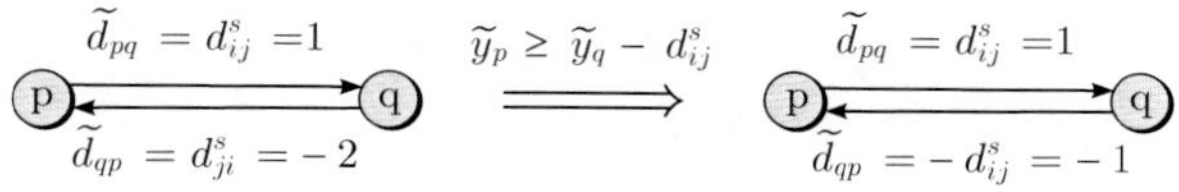

$\widetilde{d}_{pq} = d^s_{ij} = 1$, $\widetilde{d}_{qp} = d^s_{ji} = -2$ $\quad \widetilde{y}_p \geq \widetilde{y}_q - d^s_{ij} \Longrightarrow \quad$ $\widetilde{d}_{pq} = d^s_{ij} = 1$, $\widetilde{d}_{qp} = -d^s_{ij} = -1$

Fig. 3.5 For $i, j \in V$ with a minimum time lag $(i, j) \in E^+$ of duration 1 and a maximum time lag $(j, i) \in E^-$ of duration 2 in scenario $s \in S$, the *left chart* visualizes the corresponding subgraph of $\widetilde{G}$ with $p = sn + i$ and $q = sn + j$. Conducting the fixation (i, j, s) replaces the value of precedence $(q, p) \in \widetilde{E}$ as shown on the *right side*

Summing up, $\widetilde{E}$ and $\widetilde{d}$ are defined as follows:

$$\begin{aligned}\widetilde{E} = \{(0,i) \,:\, i = 1,\dots,n\} &\cup \{(i, sn+i) \,:\, i \in V,\ s \in S\} \\ &\cup \{(sn+i, i) \,:\, (i,i,s) \in \tau\} \cup \{(sn+i, sn+j) \,:\, (i,j) \in E,\ s \in S\} \\ &\cup \{(sn+j, sn+i) \,:\, (i,j,s) \in \tau,\ i \neq j\} \cup \{(sn+n, 0) \,:\, s \in S\},\end{aligned} \tag{3.9a}$$

$$\widetilde{d}_{pq} = \begin{cases} d^s_{ij} & \text{if } p = sn+i,\ q = sn+j,\ (i,j) \in E \text{ and } (j,i,s) \notin \tau, \\ -d^s_{ji} & \text{if } p = sn+i,\ q = sn+j,\ (j,i) \in E \text{ and } (j,i,s) \in \tau, \\ -\Delta & \text{if } p = sn+n,\ q = 0 \text{ and } s \in S, \\ 0 & \text{otherwise.} \end{cases} \tag{3.9b}$$

Note that if both an ordinary precedence (inherited from Υ) and a fixation exist between two nodes in $\widetilde{V}$, then the latter constraint must be more restrictive. Hence, the definition of $\widetilde{d}$ ignores the precedence from Υ in such cases.

Example 3.4.2. Figure 3.4 visualizes the construction of $\Gamma(\tau)$ for a small example network and $\tau_0 = \emptyset$ (the root node of $\mathcal{T}$). On the right side of the figure, 0 represents the artificial source, $1, \dots, 4$ the TPT nodes and $5, \dots, 12$ the task-scenario nodes. The precedences $(8, 0)$ and $(12, 0)$ enforce the deadline Δ. Figure 3.5 illustrates the fixation process.

The following theorem establishes the link between $\Upsilon(\tau)$ and $\Gamma(\tau)$.

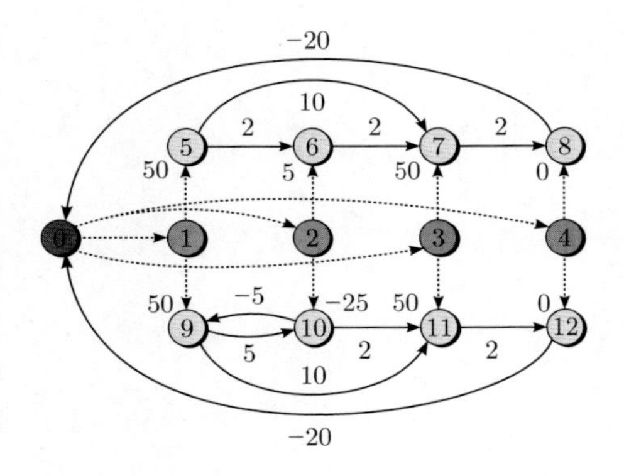

Fig. 3.6 Deterministic NPV maximization problem $\Gamma(\tau_1)$ associated with the branch-and-bound node τ_1 in Example 3.4.1

Theorem 3.4.3. *Consider a problem instance* $\Upsilon = (G, S, p, \zeta, d, \Delta, \beta)$, *a tree node* $\tau \in \mathcal{T}$ *and the deterministic NPV maximization problem* $\Gamma(\tau) = (\widetilde{G}, \widetilde{\zeta}, \widetilde{d}, \beta)$ *defined by (3.8) and (3.9). Let* $\widetilde{y}$ *be an optimal solution to* $\Gamma(\tau)$, *where* $\widetilde{y}_p$ *denotes the start time of task* $p \in \widetilde{V}$. *Then*

$$(\widetilde{y}_1, \ldots, \widetilde{y}_{\tilde{n}}) \in \arg\max \{ f(r, y) \; : \; (r, y) \in \mathcal{Y}_\tau \} .$$

Proof. Under the natural identification

$$((\widetilde{y}_1, \ldots, \widetilde{y}_n), (\widetilde{y}_{n+1}, \ldots, \widetilde{y}_{\tilde{n}})) \quad = \quad ((r_1, \ldots, r_n), (y_1, \ldots, y_{nm})) \quad = \quad (r, y),$$

the feasible sets of $\Gamma(\tau)$ and $\Upsilon(\tau)$ coincide. Furthermore, the objective value of $\widetilde{y}$ in $\Gamma(\tau)$ equals the objective value of (r, y) in $\Upsilon(\tau)$. □

Example 3.4.3. Consider the branch-and-bound node τ_1 that is generated in the solution of the stochastic NPV maximization problem in Example 3.4.1. Figure 3.6 visualizes the deterministic NPV maximization problem $\Gamma(\tau_1)$ for this node. In the figure, node 0 represents the artificial source, $1, \ldots, 4$ the TPT nodes and $5, \ldots, 12$ the task-scenario nodes. A dotted arc (p, q) refers to a precedence with duration $d_{pq} = 0$. Nodes $0, \ldots, 4$ have zero cash flows; we do not show these cash flows in the figure.

The instance $\Gamma(\tau_1)$ of the deterministic NPV maximization problem (3.2) is

$$\begin{aligned}
\underset{y}{\text{maximize}} \quad & 50 \cdot 0.95^{y_5} + 5 \cdot 0.95^{y_6} + 50 \cdot 0.95^{y_7} \\
& + 50 \cdot 0.95^{y_9} - 25 \cdot 0.95^{y_{10}} + 50 \cdot 0.95^{y_{11}} \\
\text{subject to} \quad & y \in \mathbb{R}^{13} \\
& y_1 \geq y_0, \quad y_2 \geq y_0, \quad y_3 \geq y_0, \quad y_4 \geq y_0, \\
& y_5 \geq y_1, \quad y_9 \geq y_1, \quad y_6 \geq y_2, \quad y_{10} \geq y_2, \\
& y_7 \geq y_3, \quad y_{11} \geq y_3, \quad y_8 \geq y_4, \quad y_{12} \geq y_4,
\end{aligned}$$

$$y_6 \geq y_5 + 2, \quad y_7 \geq y_5 + 10, \quad y_7 \geq y_6 + 2, \quad y_8 \geq y_7 + 2,$$

$$y_{10} \geq y_9 + 5, \quad y_{11} \geq y_9 + 10, \quad y_{11} \geq y_{10} + 2, \quad y_{12} \geq y_{11} + 2,$$

$$y_9 \geq y_{10} - 5,$$

$$y_0 \geq y_8 - 20, \quad y_0 \geq y_{12} - 20,$$

$$y_0 = 0.$$

The optimal solution to this problem is

$$(y_0, \ldots, y_{12}) = (0, 0, 0, 0, 0, 0, 2, 10, 12, 0, 5, 10, 12)$$

with objective value 115.40, see Example 3.4.1.

3.4.2 *Warm-Start Technique*

A non-root node τ' differs from its ancestor $\tau \in A_{\mathcal{T}}(\tau')$ by exactly one fixation. Hence, an optimal solution to $\Upsilon(\tau)$ is likely to be very similar to an optimal solution to $\Upsilon(\tau')$. This property carries over to the optimal solutions to the deterministic NPV maximization problems $\Gamma(\tau)$ and $\Gamma(\tau')$. This similarity of nodal solutions is typical for branch-and-bound algorithms and is exploited by warm-start techniques. In our context, this means that at node τ' we should warm-start the algorithm developed in [NZ00] (hereafter referred to as NZ) with an optimal solution to $\Gamma(\tau)$. The hope is that NZ requires significantly fewer iterations than if we apply it to a standard initial solution.

Let us elaborate this idea. Assume that $\tau' \setminus \tau = \{(i, j, s)\}$ for $(i, j) \in E^+$; the case $\tau' \setminus \tau = \{(j, j, s)\}$ for $j \in V$ is analogous. The precedence that relates i and j is more constraining in $\Gamma(\tau')$ than it is in $\Gamma(\tau)$. The modified precedence is not fulfilled by the optimal solution found for $\Gamma(\tau)$, for otherwise $(i, j, s) \notin V_\tau$ in Step 4 of the branch-and-bound algorithm and hence $(i, j, s) \notin \tau' \setminus \tau$. Since NZ is a variant of the network simplex algorithm, it has to be started from a (primal) feasible solution. If we enforced the fixation $(i, j, s) \in \tau' \setminus \tau$ as a hard constraint, then we would need to specify a feasible initial solution to $\Gamma(\tau')$. Instead, we incorporate it implicitly by penalizing its violation in the objective function:

$$g_\pi(\widetilde{y}) = g(\widetilde{y}) + \pi(\beta^{\widetilde{y}_{sn+j}} - \beta^{\widetilde{y}_{sn+i} + d^s_{ij}}).$$

Here, g denotes the objective function of problem $\Gamma(\tau)$, while g_π constitutes the penalized function for some penalty factor $\pi > 0$. Note that $\widetilde{y}_{sn+j} \geq \widetilde{y}_{sn+i} + d^s_{ij}$ holds for all feasible solutions to $\Gamma(\tau)$ and $\Gamma(\tau')$ because the respective minimum time lag is enforced in both problems. Hence, $g_\pi(\widetilde{y}) \leq g(\widetilde{y})$ for all feasible solutions $\widetilde{y}$, and $g_\pi(\widetilde{y}) = g(\widetilde{y})$ if and only if $\widetilde{y}_{sn+j} = \widetilde{y}_{sn+i} + d^s_{ij}$, that is, if

$\widetilde{y}$ obeys the new fixation $(i, j, s) \in \tau' \setminus \tau$. Note also that the penalized objective function g_π can be obtained from g by merely modifying two cash flows in $\Gamma(\tau)$:

$$\widetilde{\zeta}_{\pi,p} = \begin{cases} \widetilde{\zeta}_p - \pi\beta^{d_{ij}^s} & \text{if } p = sn + i, \\ \widetilde{\zeta}_p + \pi & \text{if } p = sn + j, \\ \widetilde{\zeta}_p & \text{otherwise.} \end{cases} \tag{3.10}$$

Hence, we can solve the penalty formulation by applying NZ to the slightly modified problem instance $\Gamma_\pi(\tau) = (\widetilde{G}, \widetilde{\zeta}_\pi, \widetilde{d}, \beta)$. The following theorem shows how this penalty formulation relates to $\Gamma(\tau')$:

Theorem 3.4.4. *Consider a problem instance* $\Upsilon = (G, S, p, \zeta, d, \Delta, \beta)$ *and* $\tau, \tau' \subseteq V^2 \times S$, *where* $\mathcal{Z}_\tau \neq \emptyset$ *and* $\tau' \setminus \tau = \{(i, j, s)\}$, $(i, j) \in E^+$. *Moreover, let* $\Gamma_\pi(\tau) = (\widetilde{G}, \widetilde{\zeta}_\pi, \widetilde{d}, \beta)$ *be the modified deterministic NPV maximization problem that penalizes the violation of* $(i, j, s) \in V_\tau$ *in the branch-and-bound node* τ. *Then there exists a penalty factor* $\pi_0 \geq 0$ *such that for all* $\pi \geq \pi_0$, *the optimal solution* $\widetilde{y}$ *to* $\Gamma_\pi(\tau)$ *found by NZ satisfies*

(i) $\widetilde{y}_{sn+j} = \widetilde{y}_{sn+i} + d_{ij}^s \iff \widetilde{y} \in \arg\max \Gamma(\tau')$;
(ii) $\widetilde{y}_{sn+j} \neq \widetilde{y}_{sn+i} + d_{ij}^s \iff \Gamma(\tau')$ *is infeasible.*

Proof. We can assume that $\mathcal{Z}_\Upsilon \cap \mathcal{Z}_{ij}^s \neq \emptyset$ since otherwise the assertion trivially holds for any $\pi_0 \geq 0$. The variable substitution presented in Sect. 3.2 transforms $\Gamma(\emptyset)$ to an equivalent LP. Being a derivate of the network simplex algorithm, NZ always terminates at a vertex of this LP, which in turn corresponds to a vertex of $\mathcal{Z}_\Upsilon$ [NZ00]. Moreover, by the nature of the hyperplane fixations, the application of NZ to $\Gamma(\tau)$ (and hence $\Gamma_\pi(\tau)$) always terminates at a vertex of $\mathcal{Z}_\Upsilon$, too. Let $\mathcal{V}$ be the finite set containing all vertices of $\mathcal{Z}_\Upsilon$ that do *not* lie on the hyperplane $\mathcal{Z}_{ij}^s$. We can assume $\mathcal{V} \neq \emptyset$ since otherwise the assertion is trivially satisfied for any $\pi_0 \geq 0$. For every $v \in \mathcal{V}$, we can determine a finite penalty factor π_v such that for all $\pi \geq \pi_v$,

$$g_\pi(v) < \min \left\{ g_\pi(\widetilde{z}) \, : \widetilde{z} \in \mathcal{Z}_\Upsilon \cap \mathcal{Z}_{ij}^s \right\}.$$

The existence of π_v follows from the fact that $\mathcal{Z}_\Upsilon$ is compact and g is bounded.

Set $\pi_0 = \max_{v \in \mathcal{V}} \pi_v$ and choose any $\pi \geq \pi_0$. If the optimal solution $\widetilde{y}$ to $\Gamma_\pi(\tau)$ lies on the hyperplane $\mathcal{Z}_{ij}^s$, then it is optimal among all elements of $\mathcal{Z}_\tau \cap \mathcal{Z}_{ij}^s$. Since g_π coincides with g on $\mathcal{Z}_{ij}^s$, $\widetilde{y}$ must then be optimal for $\Gamma(\tau')$. If, on the other hand, $\widetilde{y} \notin \mathcal{Z}_{ij}^s$, then the choice of π implies that $\mathcal{Z}_\tau \cap \mathcal{Z}_{ij}^s = \emptyset$, which is equivalent to $\mathcal{Z}_{\tau'} = \emptyset$. The reverse implications, finally, hold by definition. □

In practice, we do not need to choose π explicitly to solve $\Gamma_\pi(\tau)$. Indeed, if we employ NZ to solve $\Gamma_\pi(\tau)$, then the values of all cash flows and dual variables in the algorithm description (see [NZ00]) are of the form $\pi a + b$ for $a, b \in \mathbb{R}$. Hence, we can employ a variant of NZ that operates on tuples of cash flows and dual variables, where the tuple (a, b) corresponds to the value $\pi a + b$ for some undefined but sufficiently large π. The algorithm description from [NZ00] remains valid, the only differences being that (1) operations on cash flows and dual variables are performed entry-wise, and (2) the variable that leaves the dual basis is chosen in lexicographic order. This is reminiscent of the Big-M method in linear programming [Tah97].

Once we have obtained an optimal solution to $\Gamma_\pi(\tau)$, we can either discard the tree node τ' (in case infeasibility has been detected) or update the time lag $\widetilde{d}_{pq}$, $p = sn + j$ and $q = sn + i$, indexed by $\tau' \setminus \tau = \{(i, j, s)\}$. This allows us to use the optimal solution to $\Gamma_\pi(\tau)$ not just for the upper bound of node τ', but also as an initial solution to $\tau'' \in D_\mathcal{T}(\tau')$. The imposition of a tight maximum time lag between p and q entails that we do not require the introduced penalty terms anymore but can rather reuse the original cash flows ζ in subsequent iterations of the branch-and-bound procedure.

We close this section with an example of the warm-start procedure.

Example 3.4.4. Consider the solution to the deterministic NPV maximization problem $\Gamma(\tau_1)$ associated with the branch-and-bound node τ_1 in Example 3.4.1. In comparison to the problem $\Gamma(\tau_0)$ associated with the root node of the branch-and-bound tree, $\Gamma(\tau_1)$ contains an additional precedence $(10, 9)$ of value $\widetilde{d}_{10,9} = -5$, see Fig. 3.6. The optimal solution $\widetilde{y}$ to $\Gamma(\tau_0)$ satisfies $\widetilde{y}_9 = 0$ and $\widetilde{y}_{10} = 8$ and therefore violates the new precedence constraint $\widetilde{y}_9 \geq \widetilde{y}_{10} - 5$ contained in $\Gamma(\tau_1)$. According to Theorem 3.4.4, we can enforce this new constraint by changing the cash flows associated with tasks 9 and 10 to

$$\widetilde{\zeta}_{\pi,9} = 50 - 0.95^{-5}\pi \qquad \text{and} \qquad \widetilde{\zeta}_{\pi,10} = -25 + \pi.$$

For $\pi = 1{,}000$, for example, we obtain $\widetilde{\zeta}_{\pi,9} = -723.78$ and $\widetilde{\zeta}_{\pi,10} = 975$. This choice of cash flows guarantees that $\widetilde{y}_{10} = \widetilde{y}_9 + 5$ in any optimal solution, that is, task 10 will be started exactly 5 time units after task 9 has been started. Note that the combined cash flow of tasks 9 and 10 evaluates to

$$\widetilde{\zeta}_{\pi,9}\,\beta^{\widetilde{y}_9} + \widetilde{\zeta}_{\pi,10}\,\beta^{\widetilde{y}_{10}} = -723.78 \cdot 0.95^{\widetilde{y}_9} + 975 \cdot 0.95^{\widetilde{y}_9+5} = 30.66 \cdot 0.95^{\widetilde{y}_9}.$$

This cash flow is identical to the original combined cash flow of tasks 9 and 10 if they are started in immediate succession:

$$\widetilde{\zeta}_9\,\beta^{\widetilde{y}_9} + \widetilde{\zeta}_{10}\,\beta^{\widetilde{y}_{10}} = 50 \cdot 0.95^{\widetilde{y}_9} - 25 \cdot 0.95^{\widetilde{y}_9+5} = 30.66 \cdot 0.95^{\widetilde{y}_9}.$$

Hence, the changes in $\widetilde{\zeta}_\pi$ do not influence the task start times beyond the desired fixation.

3.5 Numerical Results

In the first part of this section, we compare TPT policies with alternative policy classes for the stochastic NPV maximization problem. In the second part, we report on the scalability of the branch-and-bound procedure and assess its performance as compared to CPLEX, a general purpose optimization package.

Table 3.1 Example temporal network with three scenarios and occurrence probabilities $p = (0.3, 0.2, 0.5)$

	Scenario 1			Scenario 2			Scenario 3		
	δ_i^1	z_i^1	ζ_i^1	δ_i^2	z_i^2	ζ_i^2	δ_i^3	z_i^3	ζ_i^3
Task 1	9	−99.6	−74.0	3	61.0	55.2	3	43.9	39.8
Task 2	5	80.7	68.4	6	145.7	119.5	4	−126.3	−110.7
Task 3	5	−136.4	−115.6	6	4.5	3.7	3	−78.1	−70.7
Task 4	7	−28.6	−22.7	8	74.6	57.3	10	172.3	123.8
Task 5	6	−32.7	−26.8	3	−92.6	−83.9	8	−37.4	−28.7

Specified are the scenario-wise task durations δ_i^s, the corresponding cash flows z_i^s at task *completion* and their discounted equivalents ζ_i^s at the task *start* times

Apart from the illustrative example at the beginning of Sect. 3.5.1, all considered test instances are randomly constructed with an adapted version of the network generator ProGen/max [NSZ03], which is known to generate difficult network instances. For the construction of the network structure, we adopt the parameter values used in the UBO instances of the PSP/max benchmark library[6] (scaled to the respective problem size). For every scenario, the task cash flows are sampled from a uniform distribution on $[-100, 100]$, while the durations of the minimum time lags are selected from a uniform distribution with support $[1, 10]$. As for the maximum time lags, let δ_{ij}^s denote the start time difference between tasks i and j in scenario $s \in S$ of the early start schedule. If the network structure (as obtained from ProGen/max) prescribes a maximum time lag between i and j, then we set its duration in scenario s to $\theta_{ij}\delta_{ij}^s$, where θ_{ij} is chosen from a uniform distribution with support $[\underline{\theta}, \overline{\theta}]$. The parameters $\underline{\theta}$ and $\overline{\theta}$ describe the tightness of maximum time lags; their values will be specified later. Similarly, we choose a value of $\theta \max_{s \in S} \Delta^s$ for the deadline, where Δ^s denotes the minimum makespan for scenario $s \in S$ and θ is sampled from a uniform distribution on $[\underline{\theta}, \overline{\theta}]$. Throughout this section, we employ a discount factor of 0.9675.

3.5.1 TPT Policies and Alternative Problem Formulations

Consider the example network encoded through the data in Table 3.1 and Fig. 3.7. In order to obtain the corresponding problem instance Υ (see Sect. 3.3), we apply the following transformations: (1) we discount the cash flows to the task start times; (2) we convert the maximum time lags to minimum time lags between the task start times; and (3) we introduce an artificial sink node. The resulting network is illustrated in Fig. 3.8.

[6] See http://www.wior.uni-karlsruhe.de/LSNeumann/Forschung/ProGenMax.

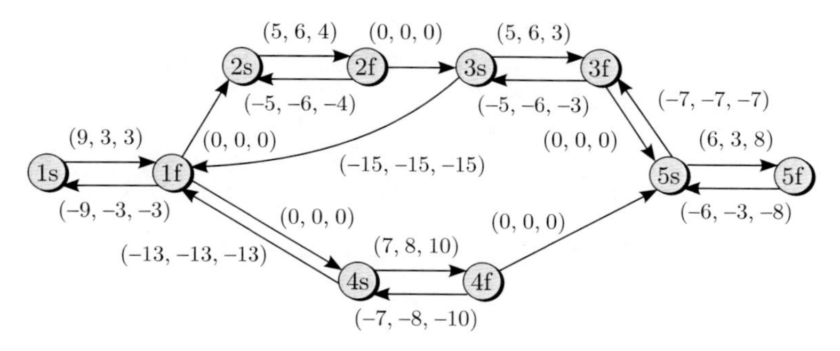

Fig. 3.7 Example temporal network in the notation of [EK92]. For $i \in V$, node is (if) represents the start (completion) event of task i. The numbers attached to an arc (i, j) describe the minimum amount of time that event i must be realized before event j in the three scenarios

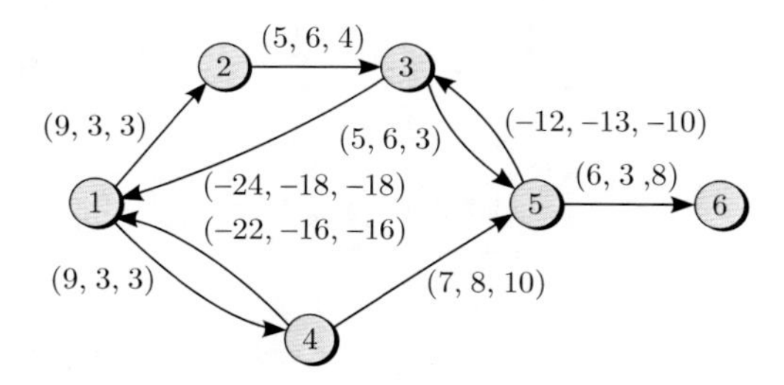

Fig. 3.8 The (standardized) problem instance Υ for the network described in Table 3.1 and Fig. 3.7. The triples of numbers attached to the arcs represent the values of the corresponding precedences in the three scenarios. The cash flow vector ζ is given in Table 3.1

For a deadline of $\Delta = 30$, the optimal TPT policy is $r^* = (0, 12, 18, 3, 22)$. Here and in the remainder of this section, we suppress the artificial sink node in the results. Policy r^* has an expected NPV of 6.25, which results from NPVs of -121.2, 151.6 and 24.6 in scenarios 1, 2 and 3, respectively. The corresponding schedules are presented in Fig. 3.9. We can identify the tendency to schedule tasks 1 and 4 early, whereas tasks 3 and 5 are delayed. This is in line with the expected cash flows of the tasks. Note that task 2 has a negative expected cash flow and should as such be scheduled late. We cannot assign a TPT larger than 12 to it, however, since otherwise the maximum time lag between tasks 1 and 3 would be violated in scenario 2.

The formulation (3.3) properly takes into account uncertainty but results in a difficult optimization problem. Hence, it is tempting to relax the computational burden by solving a simplified model to obtain a feasible schedule with an acceptable expected NPV. In the following, we compare TPT policies with three alternative approaches, namely rigid policies, nominal TPT policies and task delay policies obtained from the two-stage approach in [Ben06], see Sect. 3.2 (hereafter referred to as TD policies). Rigid policies stipulate scenario-independent task start times that satisfy the minimum and maximum time lags in all scenarios. Contrary to

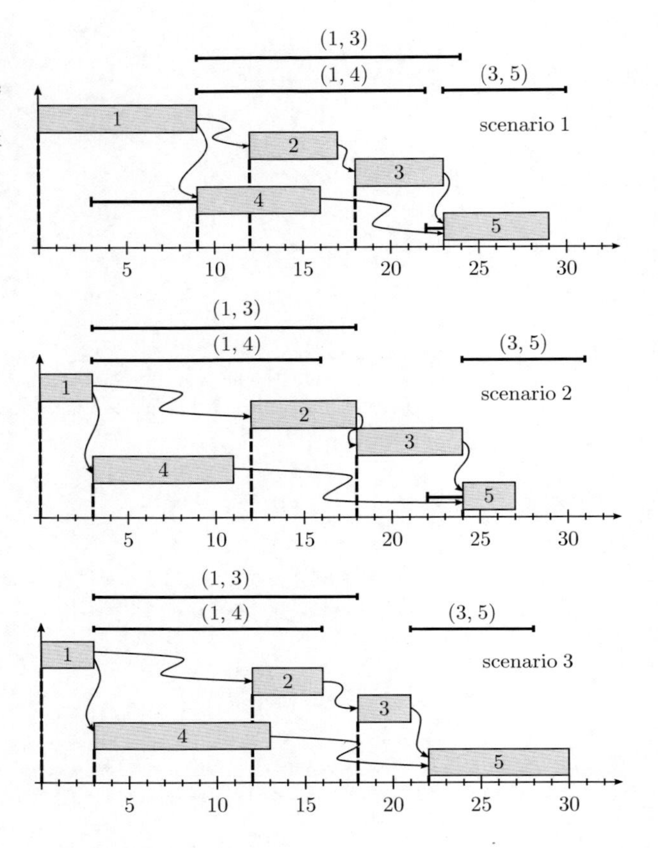

Fig. 3.9 Gantt charts for the optimal TPT policy. The *horizontal axis* represents the elapsed time, while the *vertical axis* lists the network tasks. *Arrows* between the tasks indicate finish-start precedences, whereas maximum time lags are visualized by bars above the charts

TPT policies, rigid policies never require tasks to be delayed beyond their specified start times. Optimal rigid policies can be determined by solving a deterministic NPV maximization problem which contains the time lags of all scenarios. Nominal TPT policies are obtained from a deterministic NPV maximization problem with expected values for both the uncertain time lags and the cash flows. The solution to this deterministic problem can be interpreted as a TPT policy: every task is started as early as possible, but never before its start time in the nominal solution. Even if the optimization problem which determines an optimal nominal policy is feasible, the resulting TPT policy may be infeasible due to the existence of maximum time lags and deadlines. Note that by construction, both rigid and nominal policies form subsets of the class of TPT policies, and as such they can never lead to better schedules than the optimal TPT policy determined by model (3.3). TD policies are discussed in Sect. 3.2.

For our example, the optimal rigid policy corresponds to the task start vector $(0, 10, 16, 9, 22)$ and an expected NPV of -9.5. The optimal nominal policy is $r^* = (0, 14.7, 19.4, 4.8, 23.6)$; this policy is infeasible, however, because the deadline is violated in scenarios 1 and 3, and the maximum time lag between tasks 1 and

3 is exceeded in scenarios 2 and 3. Hence, the nominal policy leads to infeasible schedules in all scenarios. The TD policy, finally, results in an expected NPV of 3.7.

Let us now determine schedules for a whole range of deadlines. Plotting the expected NPVs versus the underlying deadlines results in a curve that can be interpreted as the efficient frontier of the respective policy class. The efficient frontiers of the TPT, rigid, nominal and TD policies are shown in Fig. 3.10. The TPT schedules are feasible for all considered deadlines and outperform all other schedules. TD policies perform only slightly worse than TPT policies for deadlines below 36 time units but become infeasible for larger deadlines. This undesirable effect is caused by the approximation of a stochastic problem via a deterministic one in the two-stage approach from [Ben06] (see Sect. 3.2) and cannot occur for the TPT policies determined by model (3.3). The class of rigid policies provides feasible solutions for deadlines above 29 time units, but the resulting schedules perform substantially worse than the TPT schedules. Nominal policies, finally, yield infeasible schedules for all considered deadlines.

Since the findings from one single test instance may not be representative, we compare the performance of the aforementioned policy classes on 500 random test instances. Every instance accommodates three scenarios and ten tasks and is constructed according to the specification outlined in the beginning of Sect. 3.5 with $(\underline{\theta}, \overline{\theta}) = (1.25, 1.50)$. For the resulting test set, feasible TPT policies exist in 493 cases (98.6%). In contrast, feasible rigid policies can be determined for 258 instances (51.6%), feasible nominal policies for 148 instances (29.6%) and feasible TD policies for 303 instances (60.6%). For those cases where feasible policies have been found, Table 3.2 compares the resulting expected NPVs. It becomes apparent that optimal TPT policies outperform the other policy classes on the chosen test set. Although nominal and TD policies perform reasonably well on instances where they lead to feasible schedules, they are of limited use due to frequent infeasibilities.

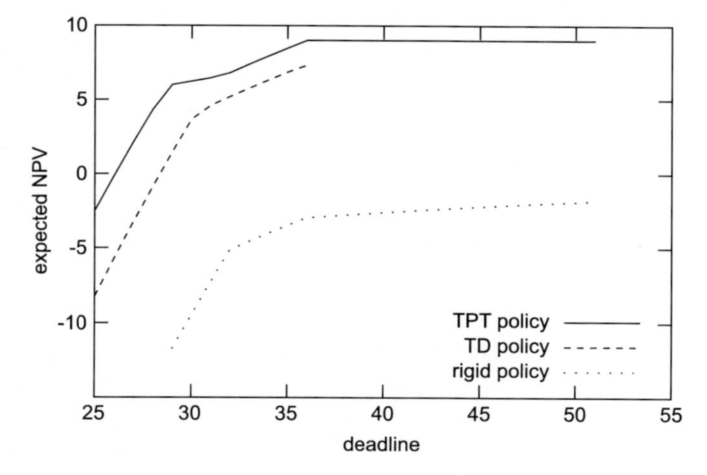

Fig. 3.10 Efficient frontiers of TPT, rigid and TD policies. Nominal policies result in infeasible schedules for all considered deadlines

Table 3.2 NPV gains of TPT policies over rigid, nominal and TD policies

	$q_{0.1}$ (%)	$q_{0.25}$ (%)	$q_{0.5}$ (%)	$q_{0.75}$ (%)	$q_{0.9}$ (%)
Rigid policies	5.35	8.93	16.02	30.24	63.66
Nominal policies	1.71	2.17	4.74	7.64	13.47
TD policies	−16.65	−3.72	2.50	23.43	49.32

The entries represent the relative increase in expected NPV when optimal TPT policies are employed instead of the policy class printed in front of the respective row. q_α denotes the α-quantile over the considered instances

3.5.2 Performance of the Branch-and-Bound Procedure

In this section, we investigate the performance of the presented branch-and-bound procedure and compare it with CPLEX 11.2, a state-of-the-art mixed-integer linear programming solver. We also analyze the change in complexity when some of the problem parameters are varied.

We first generate random test instances of problem (3.3) with ten scenarios, $(\underline{\theta}, \overline{\theta}) = (1.25, 1.5)$ and 10, 20, ..., 50 tasks (minimum time lags) according to the specification in the beginning of Sect. 3.5. For every network size, we solve 100 instances with an implementation of the branch-and-bound procedure and with CPLEX 11.2 on a quad-core Intel Xeon system with 2.33 GHz clock speed. In order to solve problem (3.3) with CPLEX, we reformulate the constraint (3.3b) via special ordered sets of type 1 [Wil99] to obtain a mixed-integer linear program. For every instance, we limit the runtime of both CPLEX and the branch-and-bound scheme to 10 min and allow an optimality gap of 1%. If an instance is not solved within this time, then the respective optimization run is considered unsuccessful and we record the incurred optimality gap. Table 3.3 summarizes the test results. As expected, larger problem instances are more difficult to solve with either method. Nevertheless, the presented branch-and-bound scheme finds optimal solutions for the majority of the test instances. In cases where an optimal solution cannot be secured, the procedure determines feasible TPT policies with moderate optimality gaps. CPLEX, on the other hand, fails to find feasible TPT policies for a large percentage of the test instances. Indeed, 10 min runtime only proves sufficient for small instances with up to 20 tasks. We conclude that the presented branch-and-bound procedure compares favorably to standard mixed-integer linear programming solvers.

We now investigate the impact of two important problem parameters, namely the number of scenarios and the tightness of the maximum time lags. To this end, we first consider test instances with 30 tasks, $(\underline{\theta}, \overline{\theta}) = (1.25, 1.5)$ and 5, 10, 20 and 30 scenarios. Table 3.4 summarizes the performance of the branch-and-bound procedure for this test set. As expected, the difficulty of problem (3.3) increases with the number of scenarios. Although the time limit is not sufficient to guarantee optimality for problem instances with 20–30 scenarios, the branch-and-bound scheme consistently determines feasible TPT policies with moderate optimality gaps. Let us now consider problem instances with 30 tasks, ten scenarios and varying values of $(\underline{\theta}, \overline{\theta})$. Table 3.5 shows that tighter maximum time lags (and

Table 3.3 Performance of the branch-and-bound algorithm and CPLEX for various instance sizes

Size	opt.	feas.	no sol.	Runtimes (s)			Optimality gaps (%)		
				$q_{0.25}$	$q_{0.5}$	$q_{0.75}$	$q_{0.25}$	$q_{0.5}$	$q_{0.75}$
10	98	2	0	0.00	0.04	0.57	1.3	1.5	1.5
	100	0	0	0.08	0.13	0.41	n/a	n/a	n/a
20	82	18	0	0.09	0.70	47.29	3.1	5.8	7.8
	79	11	10	2.20	20.20	398.96	5.5	172.8	∞
30	74	26	0	0.11	10.84	600.00	2.1	4.9	10.3
	27	12	61	256.42	600.00	600.00	∞	∞	∞
40	73	27	0	0.14	13.04	600.00	1.9	6.0	11.5
	19	6	75	600.00	600.00	600.00	∞	∞	∞
50	69	31	0	0.11	15.36	600.00	3.0	5.0	14.3
	2	2	96	600.00	600.00	600.00	∞	∞	∞

Columns 2–4 describe the numbers of instances for which optimal solutions, suboptimal but feasible solutions, and no feasible solutions have been found, respectively. We also document various quantities of the runtimes and optimality gaps. Each pair of rows describes the results of the branch-and-bound scheme (first row) and CPLEX (second row)

Table 3.4 Impact of the number of scenarios (first column) on the complexity of the problem instances

Scenarios	Opt.	Feas.	No sol.	Runtimes (s)			Optimality gaps (%)		
				$q_{0.25}$	$q_{0.5}$	$q_{0.75}$	$q_{0.25}$	$q_{0.5}$	$q_{0.75}$
5	98	2	0	0.01	0.14	1.6	4.5	4.8	4.8
10	74	26	0	0.11	10.84	600.00	2.1	4.9	10.3
20	34	66	0	58.90	600.00	600.00	3.0	6.6	21.1
30	22	77	1	600.00	600.00	600.00	4.2	7.0	20.7

All instances exhibit 30 tasks and $(\underline{\theta}, \overline{\theta}) = (1.25, 1.5)$

Table 3.5 Impact of the maximum time lag and deadline tightness $(\underline{\theta}, \overline{\theta})$ on the complexity of the problem instances

Tightness	Opt.	Feas.	No sol.	Runtimes (s)			Optimality gaps (%)		
				$q_{0.25}$	$q_{0.5}$	$q_{0.75}$	$q_{0.25}$	$q_{0.5}$	$q_{0.75}$
[1.00, 1.25]	67	33	0	0.79	21.55	600.00	2.7	5.1	13.2
[1.25, 1.50]	74	26	0	0.11	10.84	600.00	2.1	4.9	10.3
[1.50, 1.75]	78	22	0	0.09	8.62	282.06	1.8	4.9	7.9
[1.75, 2.00]	88	12	0	0.01	0.25	11.05	2.0	4.9	6.6

All instances exhibit 30 tasks and ten scenarios

deadlines) increase the difficulty of problem (3.3). Further investigations revealed that tighter maximum time lags reduce the set of feasible TPT policies, which in turn entails that the solutions $(\widehat{r}, \widehat{y})$ corresponding to the nodal upper bounds (see Step 3 of the branch-and-bound procedure) are more likely to violate the constraint (3.3b). This, however, results in a less effective pruning of the branch-and-bound tree $\mathcal{T}$ since the nodal upper bounds differ largely from the objective values of feasible

TPT policies. Nevertheless, the branch-and-bound scheme determines optimal or near-optimal TPT policies for all considered settings.

3.6 Conclusion

We presented a model for maximizing the expected NPV of a temporal network under uncertainty and discussed a branch-and-bound algorithm for its solution. We illustrated the favorable performance of the model and demonstrated the superiority of the solution algorithm over a state-of-the-art solver.

There is common agreement that in practice, NPV maximization problems in temporal networks are affected by significant uncertainty. Our tests reveal that a rigorous treatment of uncertainty is necessary in order to avoid infeasible or severely suboptimal schedules. Properly accounting for uncertainty, however, inevitably leads to computationally challenging problems, even when resource restrictions are disregarded. Thus, the results in this chapter highlight the need for suitable heuristics that allow the approximate solution of large-scale (and possibly resource constrained) problem instances.

Apart from the development of heuristic solution procedures, two promising directions for future work can be identified. Firstly, although being a popular decision criterion in the literature on temporal networks, maximizing the expected NPV seems to be in conflict with the risk aversion of decision makers. This problem can be alleviated by mapping cash flows to utilities (see Sect. 3.3), but the resulting decision criterion seems difficult to interpret. The considered model and parts of the solution procedure can be extended to maximize the conditional value-at-risk of the NPV. The conditional value-at-risk is further investigated in the next chapter. Secondly, formulating and solving the stochastic NPV maximization problem as a multi-stage recourse problem with decision-dependent structure would be of interest. Albeit intractable for realistic problem sizes, such a formulation would allow the precise quantification of suboptimality incurred from the restriction to policy classes such as TPT and task delay policies.

Chapter 4
Multi-Objective Optimization via Conditional Value-at-Risk

4.1 Introduction

In this chapter, we study an optimization problem that is defined on a temporal network and that aims to simultaneously optimize several conflicting objectives: the network's makespan, the resource costs, as well as the availability and reliability of the network. The availability and reliability of the network will be expressed as decision-dependent probabilities, whereas the network's makespan and the associated resource costs will be modeled as decision-dependent random variables. The model minimizes the conditional value-at-risk (CVaR) of the makespan and resource costs while imposing constraints on the availability and reliability of the network.

We apply the optimization problem to the service composition problem in service-oriented computing. Service-oriented computing (SOC) is a software engineering paradigm for organizing and utilizing distributed services. The central idea is to develop computer applications by assembling services, which are (potentially) located at remote sites. For an introduction to SOC, the reader is referred to [Pap03]. In recent years, SOC has gained great attention from researchers and practitioners alike. Indeed, nowadays virtually all major software vendors offer rich portfolios of SOC-related products. Although industry currently focuses on enterprise-wide solutions, it seems likely that the full benefits of SOC will only be reaped in a global service market where providers and consumers of services are dispersed among different companies, industries and locations. For this to happen, however, several theoretical and practical hurdles have to be taken first. In this chapter, we consider a particularly prominent one, namely the service composition problem.

To motivate the service composition problem, we consider a global service market. In such a market, web services cannot be composed to applications manually for at least two reasons. Firstly, the size of a global service directory quickly becomes too large for humans to process. Secondly, a global service market can be expected to be highly dynamic. On one hand, this implies that there is no

W. Wiesemann, *Optimization of Temporal Networks under Uncertainty*, Advances in Computational Management Science 10,
DOI 10.1007/978-3-642-23427-9_4,

guarantee that a web service selected at design time will still be available at runtime, or that its quality properties (such as response time) have not changed. On the other hand, even if the selected service is available at runtime, better alternatives may have emerged. In such a case, it is desirable to choose one of the alternative services instead.

As a result, computer applications should not be designed in terms of specific services (as currently supported by industrial products), but rather in terms of a formal specification of the desired application behavior. At runtime, the computer can then identify and suggest (or directly invoke) suitable services. This concept is referred to as automatic service composition and poses itself at least two interesting problems. Firstly, web services need to be described in a way that can be "understood" by computers. Secondly, algorithms need to be developed which choose appropriate services for a given a specification. In this chapter, we will concentrate on the second problem.

We define the *service composition problem* as follows: given a formal specification of an application's behavior and a set of offered services, find a service composition which satisfies the specification and maximizes the user's benefit. The specification "restructure a portfolio of stocks", for example, could be fulfilled by a composition of the services "collect historical stock data", "forecast price developments", and "optimize portfolio value", see [WGH04]. The user's benefit of a service composition is determined by an aggregation of the quality properties of the services involved. The quality of a web service, in turn, is determined through its invocation costs, response time, availability, reliability, etc. [Men02, OEH02]. Further details and a general overview of service composition can be found in [MM04].

In a global service market, where market participants do not know each other in person and contract enforcement is difficult, quality of service (QoS) becomes a crucial decision criterion. Previous optimization-based approaches to the service composition problem treat the QoS as a deterministic quantity [ZBN$^+$04, AP05, GYTZ06]. However, quality criteria such as the response time and the invocation price depend on the demand for the respective service. Given a constant amount of computational resources, the response time that a service provider can guarantee is inversely related to the demand. At the same time, prices are positively correlated with the demand. The demand for a service, in turn, fluctuates over time. Part of this erratic behavior may be explained by seasonal patterns, but future demand remains inherently uncertain. This bears similarity to the demand for electric energy [Den99], computational resources [KC02] and other non-storable goods, which is arguably uncertain. As a result, both the response time and the costs of a service should be regarded as stochastic, and the service composition problem should be treated as a decision problem under uncertainty.

As we have discussed in Chap. 2, there are two different approaches for handling uncertainty in a quantitative decision optimization framework. Either uncertainty is expressed by probability distributions, in which case we speak about *risk*, or it is denoted by uncertainty sets around deterministic values, which leads to the notion of *ambiguity*. Both concepts of uncertainty – risk and ambiguity – lead to different formulations of the resulting optimization model. In the case of ambiguity, we obtain robust optimization formulations (Sect. 2.2.2), while in the case of risk, we obtain stochastic programs (Sect. 2.2.1) or Markov decision processes (Sect. 2.2.3). In this chapter, we study an optimization model that combines both notions of uncertainty. The concept of risk is used to describe the QoS criteria of different web services. The model aims to optimize the CVaR of the network's makespan and resource costs, both of which depend on the QoS criteria of the employed web services. The concept of ambiguity, on the other hand, is used to describe the uncertainty underlying the structure of the temporal network, which is caused by conditional branches in the computer application.

This chapter is organized as follows. Section 4.2 provides an overview of the service composition literature and relates other models to the approach presented in this chapter. Section 4.3 outlines the notation as well as the setup of the optimization model, which is presented in Sect. 4.4. Section 4.5 illustrates the optimization model by means of an example, while Sect. 4.6 investigates the scalability of the model. We conclude in Sect. 4.7.

4.2 Literature Review

Given the broad scope of the service composition problem, it is not surprising that it has attracted researchers from several different fields. In fact, the problem has been addressed by both the artificial intelligence (AI) and optimization community. The AI community treats the service composition problem as an AI planning problem and concentrates on finding a service composition which meets a specified intention. Due to the nature of the AI planning problem, the QoS of the resulting composition is either ignored completely or merely incorporated in rudimental ways. As we focus on QoS-aware service composition in this chapter, we do not consider AI techniques any further and refer the interested reader to the surveys given in [Pee05, RS04].

The optimization community, on the other hand, assumes that offered services are grouped into classes of equivalent functionality, such as "conduct identity check" or "solve optimization model". We say that services which belong to the same class are *semantically equivalent*. In a global service market, it is reasonable to assume that semantically equivalent services with varying QoS properties are offered. This can be due to different providers offering similar services, or due to a single provider that segments his market by offering standard and premium versions of the same service. Contrary to AI planning techniques, which try to find services that meet a given intention, optimization approaches assume that a workflow with well-specified tasks (such as "book flight ticket" or "find hotel") has already been constructed, either manually or by an a priori solved AI planning problem. The goal is then to assign a service to every workflow task such that a QoS-related utility function is maximized. In the following, we briefly review some of the contributions that are directly related to the approach presented in this chapter.

The seminal paper [ZBN$^+$04] has been the most influential contribution in the area of optimization-based service composition to date. Apart from a local method, which solves the assignment problem for every task individually, the authors present several mixed-integer linear programming (MILP) formulations for the deterministic and stochastic service composition problem. In both cases, the quality of a service is described by its response time, invocation costs, reputation, availability and reliability. In the stochastic model, the response time of a service is assumed to be normally distributed. The authors suggest to minimize the variance of the overall duration, which they identify as the sum of service response time variances along the "critical" workflow path. However, while concentrating on a single critical path and disregarding the durations of all other paths is a valid strategy for deterministic temporal networks, this method consistently underestimates the network's makespan under uncertainty, see Sect. 1.1 and [DH02]. Contrary to the deterministic formulation developed in [ZBN$^+$04], which constitutes the basis for many extensions such as [AP05, GYTZ06], the stochastic formulation did not receive any further interest.

It is suggested in [CCGP07a, CCGP07b] to determine randomized strategies for the service composition problem. The idea is to substitute the binary service selection variables with continuous ones and interpret these as probabilities for setting the respective variable to one. This concept is very appealing as generic MILPs constitute NP-hard problems and are notoriously difficult to solve. However, the formulations presented in [CCGP07a, CCGP07b] underestimate the overall duration of the workflow because they assume interchangeability of the maximum and expectation operator. As Sect. 1.1 shows, this can lead to arbitrarily suboptimal solutions.

The paper [HCLC05] formulates the service composition problem as a Markov decision process (Sect. 2.2.3). The authors formulate the problem for a sequential workflow and do not explicitly consider the problem of finding an optimal solution. Although Markov decision processes are well-suited for purely sequential workflows where service response times can be modeled as part of the reward function, they become computationally demanding for generic (i.e., non-sequential) workflows. In fact, in such cases the response times form the state transition probabilities, which leads to an exponential growth in states and tedious approximations in case the response times are not exponentially distributed. The interested reader is referred to [KA86, TSS06] for further information on these issues.

In [RBHJ07], the response time of a service is modeled as a random variable, too. The authors use Monte Carlo simulation to investigate the distribution of the overall duration of the network's makespan. Albeit mathematically sound, the developed approach is purely analytical, that is, it assumes that the services composition problem has already been solved.

A very promising recent research direction is to account for service invocation failures. Following the model suggested in [ZBN$^+$04], virtually all service composition models address the problem of invocation failures by requiring minimum values for the overall workflow availability and reliability. Albeit simple to formulate, this approach does not allow to recover from invocation failures, for example by restarting the same service or by invoking a different one. Recently, a growing number of publications investigates these issues, see [LKY05, WY06]. Related texts investigate the possibilities to make a workflow more robust by invoking "redundant" services, that is, particularly vulnerable workflow tasks are executed several times in parallel by different services in order to increase the success probability [JL05, KD07]. We are not aware of any methods for the automatic composition of workflows that incorporate these concepts, and the ideas are still at a conceptual stage.

We conclude that the stochastic service composition problem is challenging from a theoretical point of view. Although several attempts to tackle this problem can be found in the literature, no completely satisfying method has been proposed. In the remainder of this chapter, we present a method that aims to solve the service composition problem in a mathematically sound and scalable way.

4.3 The Service Composition Problem

A service-oriented computer application constitutes a workflow of tasks that need to be completed. Any such workflow can be described through a *flowgraph*, whose nodes represent the tasks. Unlike temporal networks, flowgraphs can accommodate various types of non-deterministic flow constructs such as loops and conditional branches (i.e., "if-then-else" statements). For a general overview of workflow constructs, the reader is referred to [vdAtHKB03]. Popular workflow languages for describing service-oriented computer applications are BPEL4WS, XLANG, WSFL and XPDL, see [vdA03]. Our treatment of service-oriented applications does not rely on any particular language to characterize the underlying workflows. In contrast, we assume that we are given a finite set L of *execution flows*. Any execution flow is determined through a set of tasks executed at a specific instantiation of the workflow at hand, together with a set of precedence relations between those tasks. Thus, a particular execution flow can be obtained from the underlying flowgraph by selecting a particular branch for every if-then-else statement and a particular number of iterations for every loop. Unlike the flowgraph, the corresponding execution flows constitute temporal networks in the sense of Sect. 1.1, that is, they can be represented by directed acyclic graphs $G^l = (V^l, E^l)$, $l \in L$. The nodes of G^l, which we denote by $V^l = \{1, \ldots, n_l\}$, represent the tasks to be completed if the lth execution flow is realized. For the sake of notational simplicity, we assume that node 1 is the unique start node and node n_l the unique terminal node of the network G^l. The arcs $E^l \subseteq V^l \times V^l$, on the other hand, represent the finish-start precedence relations between the tasks in G^l, that is, $(i, j) \in E^l$ imposes the constraint that task j cannot be executed before task i has terminated. We say that the execution flow G^l is completed as soon as all tasks $i \in V^l$ are completed. More information about workflows and the corresponding execution flows is provided in Sect. 4.5 and [ZBN$^+$04].

In the sequel, we fix a specific workflow and assume that the corresponding execution flows G^l, $l \in L$, are known. We let $V = \bigcup_{l \in L} V^l$ be the collection of all tasks in the different execution flows. We assume that for every task $i \in V$ we can choose among finitely many services $s \in S(i)$ that are suited for accomplishing task i. We characterize the quality of a service by the following four criteria:

1. *Response time* t_{is}: The amount of time needed to complete task i by means of service s. The parameter t_{is} is random as it depends on the demand for the corresponding service.
2. *Invocation costs* c_{is}: The price of service s when used for task i. The parameter c_{is} is random since service providers try to smooth out demand fluctuations by adjusting prices.
3. *Availability* a_{is}: The probability that service s is available for addressing task i.
4. *Reliability* r_{is}: The probability that task i is satisfactorily fulfilled if service s is used.

We emphasize that the approach presented in this chapter is not restricted to these specific quality criteria. We assume that the response times and invocation costs are integrable random variables and thus have finite expectations. Formally, we stipulate that these random variables are defined on a probability space $(\Xi, \mathcal{F}, \mathbb{P})$, where Ξ represents the sample space of outcomes, $\mathcal{F}$ denotes a σ-algebra of subsets of Ξ, and $\mathbb{P}$ is a probability measure on $\mathcal{F}$. We then regard the response times and invocation costs as measurable functions $t_{is}, c_{is} : \Xi \mapsto \mathbb{R}_+$, $i \in V$ and $s \in S(i)$. We assume that the distribution $\mathbb{P}$ underlying the responses times and invocation costs is known. In fact, distributional information can be obtained from previous service invocations or, if available, from historical market data.

Note that the availability (reliability) of service s for task i is *not* modeled as a random variable, but as a probability a_{is} (r_{is}). This probability is related to a Bernoulli random variable that takes the value 1 with probability a_{is} (r_{is}) and 0 otherwise. The former outcome indicates that service s is available (reliable) for task i, and the latter outcome indicates that it is not.

4.4 Mathematical Programming Formulation

Let us introduce binary decision variables x_{is} for $i \in V$ and $s \in S(i)$ with the interpretation that $x_{is} = 1$ if service s is chosen for task i and $x_{is} = 0$ otherwise. In the terminology of stochastic programming (Sect. 2.2.1), the variables x_{is} constitute here-and-now decisions that must be taken before the uncertain response times and invocations costs are known. The variables x_{is} must satisfy the constraints

$$\sum_{s \in S(i)} x_{is} = 1 \qquad \forall i \in V, \tag{4.1}$$

which ensure that exactly one service is chosen for each task. For the further argumentation, fix an execution flow $l \in L$ and let $y_i^l : \Xi \mapsto \mathbb{R}_+$ denote the random time point at which task $i \in V^l$ is initiated in the execution flow l. In the terminology of stochastic programming, the variables y_i^l represent wait-and-see decisions that can be taken under the knowledge of the uncertain response times and invocation costs. By convention, the workflow execution begins at time 0, and the start times of the individual tasks must satisfy the precedence constraints stipulated by the execution flows G^l, $l \in L$. Hence, for each $l \in L$ we require that

$$y_i^l(\xi) \geq 0 \qquad \forall i \in V^l, \tag{4.2a}$$

$$y_j^l(\xi) \geq y_i^l(\xi) + \sum_{s \in S(i)} t_{is}(\xi)\, x_{is} \qquad \forall (i,j) \in E^l. \tag{4.2b}$$

Here and in the following, all constraints containing realizations $\xi \in \Xi$ of the uncertain problem parameters are understood to hold $\mathbb{P}$-almost surely. Note that y_i^l is chosen after *all* uncertain parameters are revealed, which seems to violate non-anticipativity [KW94, Pré95, RS03]: the uncertain response times and invocation costs are revealed gradually when tasks are completed, and y_i^l must only depend on information that is available at the time when task $i \in V^l$ is started in execution flow $l \in L$. We will come back to this point later in this section, see Remark 4.4.1.

In the following, we denote the overall workflow duration and the total service invocation costs for the lth execution flow by $t^l : \Xi \mapsto \mathbb{R}_+$ and $c^l : \Xi \mapsto \mathbb{R}_+$, respectively, where

$$t^l(\xi) = y_{n_l}^l(\xi) + \sum_{s \in S(i)} t_{n_l s}(\xi)\, x_{n_l s},$$

$$c^l(\xi) = \sum_{i \in V^l} \sum_{s \in S(i)} c_{is}(\xi)\, x_{is}.$$

Let us now assume that the decision maker uses the conditional value-at-risk (CVaR) to quantify the risks associated with the time and cost uncertainties. As discussed in Sect. 2.2.1, the β-CVaR of a random variable $z(\xi)$ represents the expected value of $z(\xi)$ under the assumption that $z(\xi)$ exceeds its β-VaR, that is, under the assumption that $z(\xi)$ is among the $(1-\beta)\cdot 100\%$ "worst" outcomes. Rockafellar and Uryasev have shown in [RU00] that the β-CVaR of $z(\xi)$ is equivalent to

$$\inf_{\alpha \in \mathbb{R}} \left\{ \alpha + \frac{1}{1-\beta} \mathbb{E}\left[z(\xi) - \alpha\right]^+ \right\},$$

where $[x]^+ = \max\{x, 0\}$. Hence, we can define the time and cost risk functionals for all $l \in L$ as follows:

$$R_t^l(x,y) = \mathrm{CVaR}_{\beta_t}\left[t^l(\xi)\right] = \inf_{\alpha \in \mathbb{R}} \left\{ \alpha + \frac{1}{1-\beta_t} \mathbb{E}\left[t^l(\xi) - \alpha\right]^+ \right\}, \tag{4.3a}$$

$$R_c^l(x,y) = \mathrm{CVaR}_{\beta_c}\left[c^l(\xi)\right] = \inf_{\alpha \in \mathbb{R}} \left\{ \alpha + \frac{1}{1-\beta_c} \mathbb{E}\left[c^l(\xi) - \alpha\right]^+ \right\}. \tag{4.3b}$$

The confidence levels $\beta_t, \beta_c \in [0, 1]$ are constants that reflect the decision maker's risk tolerance. The decision maker is said to be *risk-neutral* if $\beta_t = \beta_c = 0$. In this case, the risk functionals reduce to simple expectation values. By focussing only on expected execution times and costs, however, one generally fails to hedge against negative outcomes; see the discussion in Sects. 3.3 and 4.5. On the other hand, the decision maker is said to be completely *risk-averse* if β_t and β_c are close to 1. In this case, the risk functionals converge to the worst-case realizations of the execution time and costs, respectively. Optimization problems that minimize worst-case objectives are discussed in more detail in Chap. 6.

Note that at the time when the here-and-now decisions x_{is} are selected, we do not know which execution flow G^l will be realized. Moreover, it is frequently impossible to assign probabilities to the different execution flows. Hence, we assume that the execution flow is *ambiguous*, and we aim to hedge against the worst possible candidate execution flow. The worst-case risk functionals corresponding to the overall execution time and the total service invocation costs are thus defined as

$$R_t(x, y) = \max_{l \in L} \left\{ R_t^l(x, y) \right\},$$

$$R_c(x, y) = \max_{l \in L} \left\{ R_c^l(x, y) \right\}.$$

Next, we discuss the availability and reliability QoS criteria. The probability that all services corresponding to x_{is} are available in the lth execution flow amounts to

$$P_a^l(x) = \prod_{i \in V^l} \sum_{s \in S(i)} a_{is} \, x_{is} = \prod_{i \in V^l} \prod_{s \in S(i)} (a_{is})^{x_{is}}.$$

Here we have used the fact that for $\alpha_i \in \mathbb{R}$ and $z_i \in \{0, 1\}$, $\sum_i \alpha_i z_i = \prod_i (\alpha_i)^{z_i}$ if $\sum_i z_i = 1$. From the rightmost expression we can see that $P_a^l(x)$ is a log-linear function. This property will help us to simplify our optimization models below. Furthermore, the probability that all services selected by x_{is} are satisfactorily completed in the lth execution flow is given by

$$P_r^l(x) = \prod_{i \in V^l} \sum_{s \in S(i)} r_{is} \, x_{is} = \prod_{i \in V^l} \prod_{s \in S(i)} (r_{is})^{x_{is}},$$

which is also a log-linear function. Recall that the execution flows are assumed to be ambiguous. Thus, the worst-case probabilities that all chosen services are available or reliable, respectively, are obtained by taking the minimum over all possible execution flows, that is, by

$$P_a(x) = \min_{l \in L} P_a^l(x),$$

$$P_r(x) = \min_{l \in L} P_r^l(x).$$

By using the above definitions, the service composition problem can now be formulated as a mathematical optimization problem over all admissible service selections x_{is} and task start times y_i^l as follows:

$$\begin{array}{ll} \underset{x,y}{\text{minimize}} & \{R_t(x,y),\ R_c(x,y)\} \\ \text{subject to} & x_{is} \in \{0,1\},\ i \in V,\ s \in S(i), \\ & y_i^l : \Xi \mapsto \mathbb{R}_+^{n_l},\ i \in V,\ l \in L, \\ & P_a(x) \geq p_a,\quad P_r(x) \geq p_r, \\ & \text{and (4.1), (4.2).} \end{array} \tag{4.4}$$

Problem (4.4) constitutes a multi-objective decision problem that attempts to minimize two conflicting criteria simultaneously: the worst-case risk functionals for the overall execution time and for the total service invocation costs. In this model, availability and reliability are restricted to exceed the prescribed tolerance levels p_a and p_r, respectively.

Remark 4.4.1 (Non-anticipativity). In problem (4.4), the task start times y_i^l are chosen after *all* uncertain parameters are revealed. This seems to violate non-anticipativity [KW94, Pré95, RS03], which requires that y_i^l must only depend on information that is available at the time when task $i \in V^l$ is started in execution flow $l \in L$. However, one can readily show that for a fixed service assignment x_{is} and a given realization $\xi \in \Xi$ of the uncertain parameters, the constraint set (4.2) is satisfied by some functions $y_i^l(\xi)$ if and only if it is satisfied by the early start schedule $\widehat{y}^l : \Xi \mapsto \mathbb{R}_+^{n_l}$ defined through $\widehat{y}_1^l(\xi) = 0$ and

$$\widehat{y}_j^l(\xi) = \max_{i \in V^l} \left\{ \widehat{y}_i^l(\xi) + \sum_{s \in S(i)} t_{is}(\xi)\, x_{is} \;:\; (i,j) \in E^l \right\} \qquad \forall j \in V^l \setminus \{1\}.$$

The early start schedule $\widehat{y}^l(\xi)$ is well defined since the network G^l is assumed to be acyclic. Moreover, the early start schedule is non-anticipative since the task start times only depend on the completion times of predecessor tasks. The risk functional $R_t(x,y)$ is a non-decreasing function of $t^l(\xi)$, $l \in L$, and $R_c(x,y)$, $P_a(x)$ and $P_r(x)$ do not depend on the task start times $y_i^l(\xi)$. Hence, if a feasible solution $y_i^l(\xi)$ to (4.2) is anticipative, then we can replace it with the corresponding (non-anticipative) early start schedule $\widehat{y}_i^l(\xi)$ without sacrificing optimality.

As any multi-objective optimization problem, the service composition problem (4.4) does not have a unique solution. Instead, it has a family of *Pareto-optimal* solutions: a feasible solution to (4.4) is said to be Pareto-optimal if there is no other feasible solution that performs equally well or better with respect to one objective and strictly better with respect to the other. Formally speaking, a feasible solution (x^*, y^*) to (4.4) is Pareto-optimal if there is no (x, y) feasible in (4.4) with

$$R_t(x^*, y^*) \geq R_t(x, y) \quad \text{and} \quad R_c(x^*, y^*) \geq R_c(x, y)$$

such that at least one of these two inequalities is strict. The set of Pareto-optimal solutions defines an efficient frontier in the R_t–R_c plane. It can be found, for instance, by solving the parametric time risk minimization problem

$$\begin{aligned} \underset{x,y}{\text{minimize}} \quad & R_t(x, y) \\ \text{subject to} \quad & x_{is} \in \{0, 1\},\ i \in V,\ s \in S(i), \\ & y_i^l : \Xi \mapsto \mathbb{R}_+^{n_l},\ i \in V,\ l \in L, \\ & R_c(x, y) \leq \gamma, \\ & P_a(x) \geq p_a, \quad P_r(x) \geq p_r, \\ & \text{and (4.1), (4.2).} \end{aligned} \qquad (\mathcal{P}_t(\gamma))$$

If we solve $\mathcal{P}_t(\gamma)$ for different values of γ and plot the graph of the mapping $\gamma \mapsto \min \mathcal{P}_t(\gamma)$, then we obtain the set of Pareto-optimal solutions for problem (4.4). Equivalently, we can find the set of Pareto-optimal solutions for problem (4.4) by solving the following problem for different values of γ:

$$\begin{aligned} \underset{x,y}{\text{minimize}} \quad & R_c(x, y) \\ \text{subject to} \quad & x_{is} \in \{0, 1\},\ i \in V,\ s \in S(i), \\ & y_i^l : \Xi \mapsto \mathbb{R}_+^{n_l},\ i \in V,\ l \in L, \\ & R_t(x, y) \leq \gamma, \\ & P_a(x) \geq p_a, \quad P_r(x) \geq p_r, \\ & \text{and (4.1), (4.2).} \end{aligned} \qquad (\mathcal{P}_c(\gamma))$$

At this stage, we should mention that the multi-objective optimization problem (4.4) and the equivalent reformulations $\mathcal{P}_t(\gamma)$ and $\mathcal{P}_c(\gamma)$ may be too conservative for decision makers with a high risk tolerance. Instead of using pessimistic worst-case risk measures for the candidate execution flows, these decision makers could use the following averaged risk measures:

$$\overline{R}_t(x, y) = \sum_{l \in L} \varrho^l R_t^l(x, y), \qquad \overline{P}_a(x) = \sum_{l \in L} \varrho^l P_a^l(x, y),$$

$$\overline{R}_c(x, y) = \sum_{l \in L} \varrho^l R_c^l(x, y), \qquad \overline{P}_r(x) = \sum_{l \in L} \varrho^l P_r^l(x, y),$$

where ϱ^l, $l \in L$, are positive weights that sum up to one. Replacing the worst-case risk measures with their averaged counterparts in problem (4.4) yields a new multi-objective optimization problem, and the corresponding parametric optimization problems $\overline{\mathcal{P}}_t(\gamma)$ and $\overline{\mathcal{P}}_c(\gamma)$ are obtained in the obvious way. In the following, we restrict ourselves to the robust problem (4.4). Nevertheless, all conclusions of this chapter are also valid for the models obtained from replacing the worst-case risk measures by their averaged counterparts. In particular, the computational complexity of $\mathcal{P}_t(\gamma)$, $\mathcal{P}_c(\gamma)$, $\overline{\mathcal{P}}_t(\gamma)$ and $\overline{\mathcal{P}}_c(\gamma)$ is similar.

All numerical calculations presented below are based on the problem $\mathcal{P}_t(\gamma)$. In the remainder of this section, we show how this problem can be approximated by a mixed-integer linear program (MILP) that can be solved via standard optimization techniques. By using representation (4.3) for the risk functionals and by taking the logarithm of the equations involving $P_a^l(x)$ and $P_r^l(x)$, problem $\mathcal{P}(\gamma)$ can be reformulated as follows:

$$
\begin{aligned}
\underset{x,y,z,t_+^l,c_+^l,\alpha_t^l,\alpha_c^l}{\text{minimize}} \quad & z \\
\text{subject to} \quad & x_{is} \in \{0,1\}, \ i \in V, \ s \in S(i), \\
& y_i^l : \Xi \mapsto \mathbb{R}_+, \ i \in V, \ l \in L, \\
& t_+^l, c_+^l : \Xi \mapsto \mathbb{R}_+, \ l \in L, \\
& z \in \mathbb{R}, \ \ \alpha_t^l, \alpha_c^l \in \mathbb{R}_+, \ l \in L, \\
& \alpha_t^l + (1-\beta_t)^{-1}\mathbb{E}\left[t_+^l(\xi)\right] \le z, \\
& \alpha_c^l + (1-\beta_c)^{-1}\mathbb{E}\left[c_+^l(\xi)\right] \le \gamma, \\
& t_+^l(\xi) \ge y_{n_l}^l(\xi) + \sum_{s\in S(i)} t_{n_l s}(\xi)\, x_{n_l s} - \alpha_t^l, \\
& c_+^l(\xi) \ge \sum_{i\in V^l}\sum_{s\in S(i)} c_{is}(\xi)\, x_{is} - \alpha_c^l, \\
& y_j^l(\xi) \ge y_i^l(\xi) + \sum_{s\in S(i)} t_{is}(\xi)\, x_{is} \qquad \forall (i,j) \in E^l, \\
& \sum_{i\in V^l}\sum_{s\in S(i)} \ln(a_{is})\, x_{is} \ge \ln(p_a), \\
& \sum_{i\in V^l}\sum_{s\in S(i)} \ln(r_{is})\, x_{is} \ge \ln(p_r), \\
& \sum_{s\in S(i)} x_{is} = 1 \qquad \forall i \in V.
\end{aligned}
\tag{4.5}
$$

In this problem, all but the last constraints are understood to hold for all execution flows $l \in L$. The real numbers z, α_t^l, and α_c^l, as well as the integrable random variables $t_+^l(\xi)$ and $c_+^l(\xi)$, represent auxiliary free decision variables for all execution flows $l \in L$. The above reformulation of $\mathcal{P}_t(\gamma)$ is manifestly an MILP. However, if the random service execution times and costs are continuously distributed, as will be the case for our examples in Sect. 4.5, then problem (4.5) represents an infinite-dimensional MILP. Moreover, the use of the expectation operators requires multi-dimensional integration to evaluate the constraints. We avoid these complications by replacing the sample space Ξ with a finite subset of scenarios ξ^k, $k \in K$, which is obtained by sampling from the probability distribution $\mathbb{P}$. Under this approximation, any random variable $z(\xi)$ reduces to a finite set of real numbers z^k, $k \in K$, while an expectation value of the form $\mathbb{E}[z(\xi)]$ reduces to the sample average $|K|^{-1} \sum_{k \in K} z^k$. This *sample average approximation* makes problem (4.5) computationally tractable, see [WA08].

4.5 Case Study

We now investigate some of the properties of the stochastic service composition model (4.5). To this end, we compare problem (4.5) with a deterministic model that is representative for the mainstream approaches in the literature. The latter can be seen as a deterministic version of $\overline{\mathcal{P}}_t(\gamma)$, which involves only averaged quality criteria. In the following, we refer to problem (4.5) as the *risk-aware problem*, while the reference model is referred to as the *nominal problem*. Similarly, an optimal solution to the risk-aware problem is called a *risk-aware composition*, while an optimal solution to the nominal problem is referred to as a *nominal composition*. By definition, the nominal problem is obtained from problem $\overline{\mathcal{P}}_t(\gamma)$ if we replace the response times and invocation costs with their expected values and treat the task start times $y_i^l(\xi)$ as deterministic decision variables. This implies that the workflow durations $t^l(\xi)$ and the total service invocations costs $c^l(\xi)$ become deterministic, too, which renders the CVaR risk functionals redundant. In fact, the CVaR risk functionals act like identity mappings when applied to deterministic quantities.

We compare the risk-aware and the nominal model by looking at the workflow in Fig. 4.1, which is borrowed from an example in [ZBN+04]. The workflow consists of six real tasks (nodes 2–7) and two artificial tasks (nodes 1 and 8). The outgoing arcs of node 1 form an AND-split, that is, both paths emanating from node 1 have to be executed. The outgoing arcs of node 5, on the other hand, form an XOR-split, that is, exactly one of the two paths emanating from node 5 needs to be executed. Thus, we obtain two execution flows corresponding to the two choices in the XOR-split, see Fig. 4.1. We assume that two candidate services are offered for every non-artificial task: a "standard" service which is cheap but suffers from high response time variability, and a more expensive "premium" service

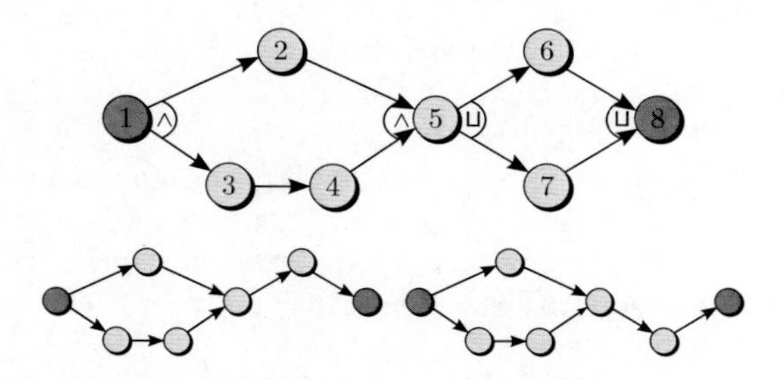

Fig. 4.1 Example workflow (upper chart) and its two execution flows (lower charts). All paths emanating from node 1 (AND-split) and exactly one path emanating from node 5 (XOR-split) have to be executed in the workflow.

Table 4.1 Distributional parameters for the workflow in Fig. 4.1

Task	Service	Response times			Invocation costs		
		opt.	m.l.	pess.	opt.	m.l.	pess.
2	Standard	4.06	8.11	16.22	1.63	1.81	2.26
	Premium	7.30	8.11	9.33	2.04	2.26	2.83
3	Standard	1.49	2.98	5.96	2.93	3.25	4.06
	Premium	2.68	2.98	3.43	3.66	4.06	5.08
4	Standard	2.93	5.86	11.72	7.70	8.56	10.70
	Premium	5.27	5.86	6.74	9.63	10.70	13.38
5	Standard	0.50	1.00	2.00	4.79	5.32	6.65
	Premium	0.90	1.00	1.15	5.99	6.65	8.31
6	Standard	1.40	2.80	5.60	8.35	9.28	11.60
	Premium	2.52	2.80	3.22	10.44	11.60	14.50
7	Standard	2.62	5.23	10.46	6.00	6.67	8.34
	Premium	4.71	5.23	6.01	7.50	8.34	10.42

For each parameter, we specify the most optimistic ("opt."), most likely ("m.l.") and most pessimistic ("pess.") values

whose response time is less volatile. The response times and prices are modeled as independent, beta-distributed random variables. We use beta distributions because they can be asymmetric and have fat upper tails, which have been observed in the empirical distributions of response times and service prices [RBHJ07]. Furthermore, beta distributions are ubiquitous in the related research area of project scheduling, see [DH02]. A beta distribution is uniquely determined by its most optimistic, most likely and most pessimistic values, that is, by its mode and the two extreme points of its support. Table 4.1 lists the values that we assign to these three parameters for the response time and invocation costs of every candidate service in the example. The availabilities and reliabilities of all services are set to 1 in order to isolate the effects due to uncertain response times and invocation costs.

We first set $\gamma = 34$ and solve both the risk-aware and the nominal problem. When solving the risk-aware model, we use a sample size of 100 and set $\beta_t = \beta_c = 0.95$. When solving the nominal model, we use weighting factors $\rho^1 = \rho^2 = 1/2$ for both execution flows. Since the response times and costs are random variables with given distributions, the duration and the costs of the overall workflow are random, too. Figure 4.2 shows the empirical probability density functions of the total service invocation costs associated with the risk-aware and the nominal composition, respectively, assuming that either execution flow is indeed realized with probability 1/2. We see that the realized costs are smaller than $\gamma = 34$ in more than 95% of the scenarios if the risk-aware composition is implemented. Similarly, the *expected* costs are smaller than 34 if the nominal composition is implemented. In the latter case, however, the costs exceed the target level 34 in roughly 50% of the scenarios. This phenomenon, as well as the discrepancy between the density function shapes (unimodal vs. bimodal), is due to the fact that the risk-aware composition chooses

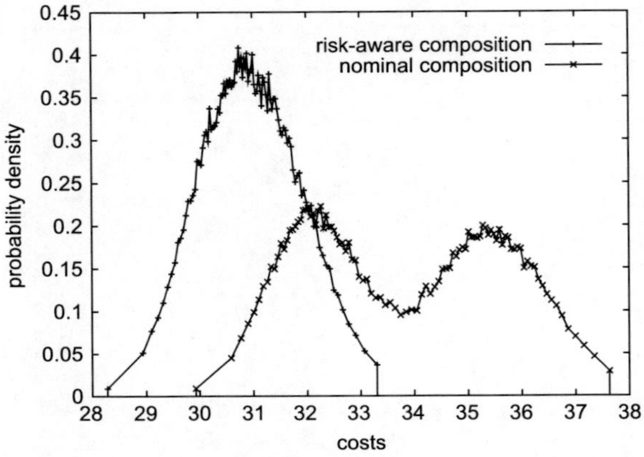

Fig. 4.2 Probability density functions of the total service invocation costs. The corresponding mean values are 31.04 (risk-aware model) and 33.79 (nominal model)

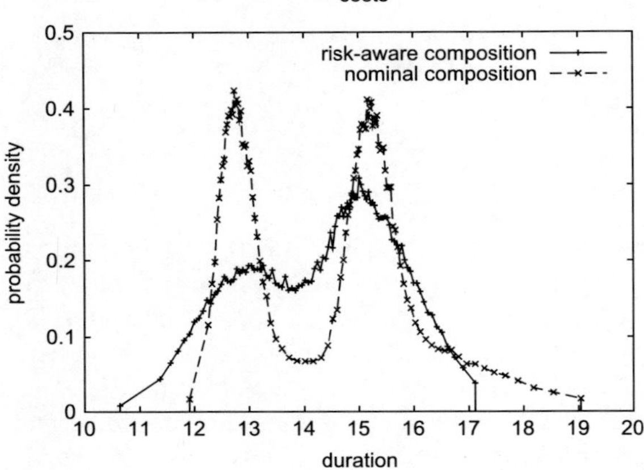

Fig. 4.3 Probability density functions of the overall workflow duration. The corresponding mean values are 14.26 (risk-aware model) and 14.59 (nominal model)

"standard" services, whereas the nominal composition chooses "premium" services for tasks 5 and 6. The high price variability of the "premium" service for task 6 leads to high costs if the second execution flow is realized. As the nominal problem accounts for the expected values only, the choice of "premium" services for tasks 5 and 6 is perfectly understandable.

Figure 4.3 shows the empirical probability density functions of the overall workflow durations of the risk-aware and the nominal compositions. We again see that the risk-aware composition performs significantly better than the nominal one in view of unfavorable outcomes. In summary, we find that the risk-aware composition is far less likely to yield high costs or long execution times than the nominal composition. Due to the correct treatment of the underlying randomness, the risk-aware composition outperforms the nominal one even in terms of expected costs and expected duration. A remarkable fact, which has been observed in the past (see, e.g., [BTN00]), is that the risk-aware solution barely sacrifices any performance in the expected case in order to gain a high degree of robustness with respect to unfavorable cases.

Up to now, we have compared the risk-aware and the nominal model only for one particular value of γ. We can obtain all Pareto-optimal compositions for both models if we regard γ as an adjustable parameter. The risks associated with these compositions are conveniently assessed via simulation, again assuming probabilities of 1/2 for either execution flow. The resulting 0.95-CVaR risk levels of the overall workflow duration and the total service invocation costs are displayed in Fig. 4.4. The graph of the mapping $\gamma \mapsto \min \mathcal{P}_t(\gamma)$ (dashed line) is interpreted as an efficient frontier. By construction, all risk-aware compositions lie on the efficient frontier, and there is no feasible (and, *a fortiori*, no nominal) composition below it. Put differently, all nominal compositions are dominated by risk-aware ones, in the sense that for every nominal composition there exists a risk-aware composition with smaller cost and duration risks. As Fig. 4.4 attests, the reduction of these risks can be significant.

In all previous tests, the sample size in the risk-aware model was fixed to 100. We now investigate the impact of increasing the sample size on the accuracy of the resulting risk-aware compositions. To this end, we again set $\gamma = 34$ and solve 500 instances of the risk-aware problem with sample size 50, 100, 150, 200 and 250. Each instance relies on a different set of samples and will thus result in a different value for both the objective function (duration risk) and the left-hand side value of the cost constraint (cost risk). Figure 4.5 visualizes the resulting cost risk estimates as a boxplot. We can see that with increasing sample size, the estimated cost risk converges to the true 0.95-CVaR of the costs associated with the true optimal composition. Similarly, Fig. 4.6 shows the estimated objective function values as a boxplot. A similar kind of convergence can be noticed. We can observe that the approximations based on small sample sizes are downward biased. This is a manifestation of the general result that any random sampling scheme provides a statistical lower bound for the true minimum objective value of a stochastic optimization problem, see [Sha03].

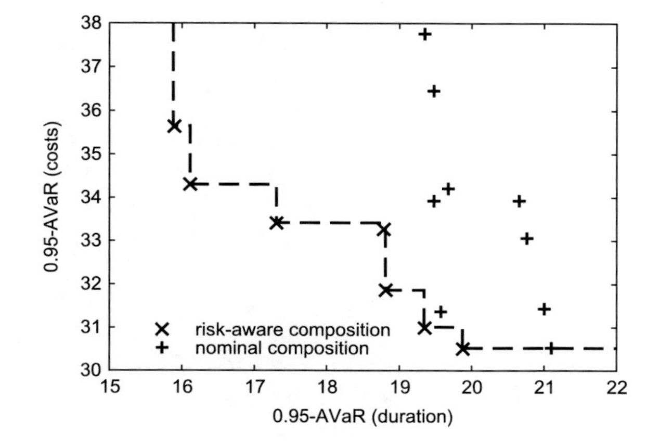

Fig. 4.4 Duration and cost risks implied by the risk-aware and nominal compositions. The *dashed line* visualizes the efficient frontier

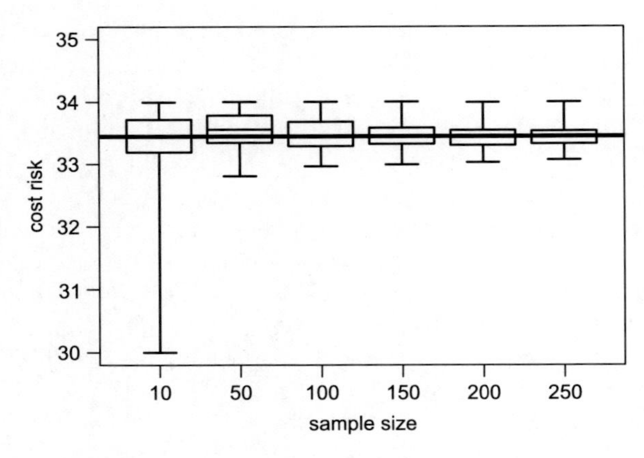

Fig. 4.5 Boxplot showing the distribution of the 95%-CVaR estimates of the total service invocation costs. The individual box-and-whisker marks represent (from bottom to top) the smallest value, 0.25-quantile, median, 0.75-quantile and the largest value, respectively. The *solid line* represents the true 0.95-CVaR of the costs

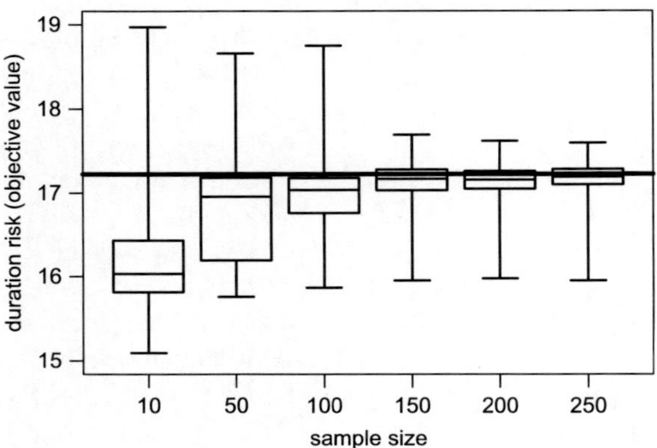

Fig. 4.6 Boxplot showing the distribution of the 95%-CVaR estimates of the overall workflow duration. The box-and-whisker marks have the same meaning as in Fig. 4.5. The *solid line* represents the true 0.95-CVaR of the duration

4.6 Scalability

In order to examine the scalability of the presented approach, we solve randomly generated test instances of the stochastic service composition problem (4.5). Every test instance consists of four different execution flows, where any two of these execution flows differ in at most 10% of their precedence relations. The algorithm to construct an execution flow is a straightforward adaptation of the one presented in [DDH93] for project scheduling problems. We again model the response times and the invocation costs of the available services as independent beta-distributed random variables. The quality requirements (costs, availability and reliability) for the overall workflow and the QoS criteria of the candidate services are chosen such that roughly 1/8 of all possible service compositions become feasible.

Figure 4.7 and Table 4.2 show the results for instances with 5, 10, ..., 50 tasks and 5, 5–10, 5–20 and 5–30 candidate services per task, respectively. We solved 100 test instances for every parameter setting. The results were obtained for an Intel Pentium 4 processor with 2.4 GHz clock speed, 4 GB main memory and the CPLEX

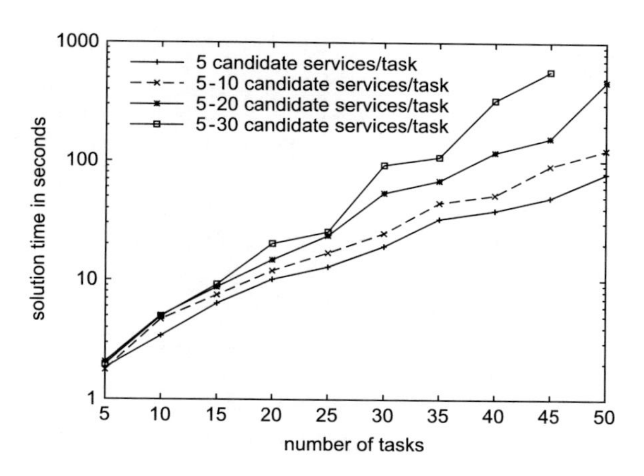

Fig. 4.7 Results of the scalability experiment. The last value for 5–30 candidate services/task is missing since the median solution time exceeds 600 s

Table 4.2 Results of the scalability experiment

Tasks	Number of candidate services							
	5		5–10		5–20		5–30	
10	100%	(n/a)	100%	(n/a)	100%	(n/a)	100%	(n/a)
20	100%	(n/a)	100%	(n/a)	100%	(n/a)	100%	(n/a)
30	100%	(n/a)	99%	(1.84%)	98%	(3.13%)	93%	(4.01%)
40	99%	(2.22%)	93%	(3.24%)	69%	(4.92%)	59%	(7.31%)
50	91%	(3.71%)	87%	(5.30%)	52%	(7.56%)	43%	(9.56%)

Shown are the percentages of solved instances (first value) and the median optimality gaps of the unsolved instances (second value; in parentheses)

8.0 optimization package. In all instances, the sample size was fixed to 100. The optimization was stopped as soon as an optimality gap of less than 1% was reached or a time limit of 600 s was exceeded. The figure reveals that the solution time increases exponentially in both the number of tasks and the number of candidate services per task. If we assume that real-life workflows do not have more than 30 tasks and 5–20 candidate services per task, then the risk-aware problem can be solved within 1–2 min. If significantly more candidate services exist for every workflow task, then the computational requirements of problem (4.5) may become unacceptable. In this case, one may try to reduce the computation time by conducting a task-wise pre-selection, which can be achieved through a local search method.

4.7 Conclusion

In practice, the precise quality of a web service is uncertain prior to its execution. Nevertheless, most of the service composition models presented in the literature are deterministic and thus require point estimates, such as expected values, for all quality measures. In this chapter, we demonstrated that such an approach may lead to suboptimal decisions. We then presented a service composition model based on stochastic programming which accounts for quality uncertainty in a mathematically sound manner. A crucial observation has been that the workflow of a computer application can be decomposed into a set of execution flows, each of which can be represented as a temporal network. The stochastic service composition model presented in this chapter allows to adjust for the risk attitude of the decision maker in a straightforward way by selecting a risk tolerance level for each quality criterion. We illustrated the favorable properties of the stochastic model through a small case study. Finally, we showed that the stochastic service composition model remains tractable for realistic problem sizes.

The work presented in this chapter can be extended in multiple directions. Firstly, it would be desirable to improve the scalability of the model. To this end, one could refine the crude Monte Carlo sampling employed in this chapter by applying various discretization schemes commonly used in stochastic programming, for example variance reduction techniques [PK05] and analytical discretization procedures [Pfl01] based on stability criteria of stochastic optimization problems [RR02]. Furthermore, due to the inherent difficulty of integer optimization problems, heuristic and approximate solution procedures may prove beneficial. Another fruitful avenue for extensions is to explicitly account for unavailable and unreliable services. So far, all optimization models that we are aware of stipulate a minimum availability and reliability, respectively, for the overall workflow. Such an approach suffers from two major shortcomings. On one hand, reoptimization is necessary whenever a service is unavailable or unreliable. On the other hand, the resulting composition is overly cautious as it does not account for recourse decisions that are available to the decision maker once a service turns out to be unavailable or unreliable. First steps towards a proper modeling of availability and reliability have been taken in [JL05, KD07, LKY05, WY06], but to the best of our knowledge, these ideas have not been incorporated in any optimization model yet.

Chapter 5
Minimization of Makespan Quantiles

5.1 Introduction

In this chapter, we consider temporal networks whose task durations are functions of a resource allocation that can be chosen by the decision maker. The goal is to find a feasible resource allocation that minimizes the network's makespan. We focus on non-renewable resources, that is, the resources are not replenished, and specified resource budgets must be met. The resource allocation model presented in this chapter is primarily suited for project scheduling problems, and for ease of exposition we will use project scheduling terminology throughout this chapter. In project scheduling, it is common to restrict attention to non-renewable resources and disregard the per-period consumption quotas that exist for renewable and doubly constrained resources, see Sect. 2.1. Apart from computational reasons, this may be justified by the fact that resource allocation decisions are often drawn at an early stage of a project's lifecycle at which the actual resource availabilities (which are unpredictable due to staff holidays, illness and other projects) are not yet known. Thus, the goal of such resource allocation models is to decide on a rough-cut plan which will be refined later.

The first resource allocation models for project scheduling have been proposed in the early 1960s. The basic model is the linear time/cost trade-off problem [Ful61, Kel61], which considers a single resource and postulates affine relationships between investment levels and activity durations. The affinity assumption implies that the marginal costs of reducing a task's duration do not depend on the current investment level. In reality, however, the marginal costs typically increase with the investment level because additional time savings are more costly to achieve (due to reliance on overtime, rented machinery, complex process changes, etc.). Indeed, linear programming theory implies that the assumption of constant marginal costs results in a pathological resource allocation behavior: the investment levels of most activities will be at one of the pre-specified investment bounds. This does not reflect reality, where prudent project managers refrain from depleting their reserves in the planning stage.

W. Wiesemann, *Optimization of Temporal Networks under Uncertainty*, Advances in Computational Management Science 10, DOI 10.1007/978-3-642-23427-9_5, © Springer-Verlag Berlin Heidelberg 2012

In order to overcome this weakness, several nonlinear resource allocation models have been suggested. A single-resource model with convex quadratic relationships between investment levels and task durations is presented in [DHV+95]. The resulting quadratic program can be solved very efficiently. Furthermore, the marginal costs of reducing a task's duration are increasing, as desired. A resource allocation problem that assigns the single resource "overtime" to project tasks is formulated in [JW00]. The authors postulate an inverse-proportional relationship between a task's duration and the amount of overtime spent on that task. Furthermore, the per-period costs of overtime are assumed to be quadratic in the amount of overtime, which leads to task expenditures that are linear in the investment levels. With this choice of functions, the resulting model is convex and can be solved efficiently. Apart from these two prototypical models, several solution procedures for single-resource models have been proposed [DH02].

So far we only mentioned single-resource models. By convention, these models concentrate on the bottleneck resource within a company. In practice, however, one frequently faces situations where multiple resources (e.g., both labor and capital) are scarce and need careful rationing. Note that due to market frictions different resources (such as permanent and temporary workers) are typically not equivalent or exchangeable. Hence, a multi-resource problem cannot generally be converted to a problem with a single "canonical" resource such as capital. To the best of our knowledge, the only problem class that accounts for multiple resources is the class of discrete multi-mode problems [DH02], which also accommodates per-period consumption quotas for the resources. Multi-mode problems assume that every project task is performed in one of finitely many different execution modes, and every execution mode implies a predefined per-period consumption of every resource. Multi-mode problems are very difficult to solve due to their combinatorial nature. Firstly, it is well known that consumption quotas per unit time lead to $\mathcal{NP}$-hard "packing" problems since the early start policy (1.2) is no longer guaranteed to be feasible, see Sect. 1.1. Secondly, the number of execution modes per activity is likely to increase rapidly in the number of resources. As a result, exact solution techniques are limited to small projects, and one typically has to resort to heuristics.

In this chapter, we present a continuous resource allocation model for project scheduling. Contrary to existing continuous models, it can accommodate multiple resources. Unlike multi-mode problems, however, the resulting optimization model is convex and hence computationally tractable. The relationship between investment levels and task durations is inspired by microeconomic theory, which makes the model justifiable and amenable to economic interpretation. Note that in practice, some of the resources might be discrete (such as staff or machinery). In this case, one can either solve the model as a continuous relaxation and use randomized rounding techniques, or one can treat the respective investment levels as integer variables and solve the resulting mixed-integer nonlinear program via branch-and-bound techniques.

In practice, some of the parameters of project scheduling problems (most notably the work contents of the project tasks) are subject to a high degree of uncertainty. One way to account for this uncertainty is to minimize the expected project makespan, see Chap. 3. However, as we have discussed in that chapter, the expected value may not be an appropriate decision criterion in project scheduling due to the non-recurring nature of projects and the high risks involved. Instead, it may be better to optimize a risk measure that also accounts for the variability of the makespan.

As pointed out in Sect. 2.2.1, two risk measures have gained notable popularity: the value-at-risk (VaR) and the conditional value-at-risk (CVaR). The α-VaR of a random variable is defined as its α-quantile. For high values of α (e.g., $\alpha \geq 0.9$), minimizing the α-VaR of the project makespan leads to resource allocations that perform well in most cases. In recent years, VaR has come under criticism due to its nonconvexity, which makes the resulting optimization models difficult to solve. Moreover, the nonconvexity implies that VaR is not sub-additive and hence not a coherent risk measure in the sense of [ADEH99]. Finally, VaR only refers to a particular quantile of a random variable but does not quantify the degree by which that quantile is exceeded "on average", if it is exceeded. All three shortcomings are rectified by CVaR. Roughly speaking, the α-CVaR of a random variable is defined as the expected value of its $(1-\alpha) * 100\%$ "worst" possible realizations, see Sect. 2.2.1 and Chap. 4. Contrary to VaR, CVaR is a coherent and, a fortiori, convex risk measure, which makes it attractive for optimization models. In the context of project scheduling, however, the advantages of CVaR over VaR seem less clear. Firstly, although the exact optimization of the α-VaR is indeed difficult, we will see in Sects. 5.3 and 5.4 that we can efficiently approximate this value with high precision. Furthermore, although being a convex risk measure, there is usually no "attractive" closed-form expression for the CVaR, and one has to rely on costly approximation or bounding techniques. Secondly, in the context of project scheduling it is not obvious why a risk measure should be sub-additive. In a financial context, sub-additivity relates the risk of individual asset portfolios to the risk of their combination. Sub-additivity becomes more difficult to interpret in the context of managing an *individual* project, however, since such a project cannot be combined with others to form a project portfolio. Whether a quantification of the risk beyond a certain quantile of the project makespan is desirable, finally, depends strongly on the contractual agreements between the project partners. For an overview of stochastic programming-based project scheduling techniques, see [HL05].

A popular alternative to the optimization of VaR and CVaR is robust optimization, see Sect. 2.2.2 and Chap. 6. Since robust optimization in its "classical" form evaluates solutions in view of their worst-case performance, it can lead to very cautious decisions. To alleviate this problem, robust optimization has been extended to incorporate distributional information about the random variables [CSS07]. Since only partial knowledge is required about the distributions of the underlying random variables, this is particularly attractive for applications in which distributions are difficult to estimate. However, this comes at the cost of rather weak approximations of the real distributions in common cases. Indeed, as we will see in Sects. 5.3 and 5.4, the use of robust optimization techniques can result in a significant overestimation of the uncertain makespan of a project under commonly accepted distributional assumptions.

As part of this chapter, we extend the deterministic multi-resource allocation model to the case of parameter uncertainty. We consider a two-stage chance constrained problem in which the resource allocation is chosen here-and-now, whereas the task start times are modeled as a wait-and-see decision, see Sect. 2.2.1. We assume that the first and second moments of the uncertain parameters are known, and we minimize an approximation of the α-VaR of the project makespan. We also present a generalization of the stochastic resource allocation model that accommodates imprecise knowledge about the moments. Contrary to the stochastic resource allocation models commonly found in the literature, the approach presented here utilizes normal approximations of the task path durations. This allows to employ a scenario-free approach which scales favorably with the problem size. At the same time, we will see that normal approximations describe the uncertain makespan significantly better than some of the bounds that are commonly used in robust optimization. Normal approximations of task paths have been first suggested for analyzing project makespans [DH02]. Recently, they have been used to obtain bounds for "risk-adjusted" deterministic circuit design [KBY$^+$07]. The use of normal approximations in the *optimization* of temporal networks has first been proposed in [WKRa]. Although we present the VaR approximation in the context of project scheduling, the formulation readily applies to other application areas of temporal networks (e.g., the design of digital circuits and the handling of production processes) as well.

The remainder of this chapter is organized as follows. In the next section we present the deterministic resource allocation model. In Sect. 5.3 we assume that some of the problem parameters are random, and we minimize an approximation of the α-VaR of the project makespan. Section 5.4 provides numerical results. In Sect. 5.5 we illustrate how we can accommodate imprecise moment information. We also discuss the iterative solution of the stochastic resource allocation model based on semi-infinite programming principles. We conclude in Sect. 5.6.

5.2 Deterministic Resource Allocation

We define a project as a temporal network $G = (V, E)$ whose nodes $V = \{1, \dots, n\}$ denote the activities (e.g., "conduct market research" or "develop prototype") and whose arcs $E \subseteq V \times V$ denote the temporal precedences among the activities in finish-start notation, see Sect. 1.1. Our goal is to find an optimal resource allocation $x \in \mathbb{R}^{mn}_{+}$, where x^k_i denotes the amount of resource $k \in K = \{1, \dots, m\}$ assigned to activity $i \in V$. Typical project resources are capital and different categories of labor and machinery. Admissible resource allocations must satisfy process and budget constraints. We assume that the *process constraints* are of box type, $\underline{c} \leq x \leq \overline{c}$, where $\underline{c}$ and $\overline{c}$ are given vectors in $\mathbb{R}^{mn}_{+}$. The components $\underline{c}^k_i$ and $\overline{c}^k_i$ denote the minimal and maximal investment levels of resource k in activity i, respectively. The budget of resource k is denoted by B_k, and the *budget constraints* require that

$\sum_{i \in V} x_i^k \leq B_k$ for all $k \in K$. Note that all project resources are assumed to be non-renewable, which has an impact on the admissible units of measure. The resource "labor", for example, can be measured in terms of man-hours. This implies that higher numbers of man-hours lead to shorter activity durations, which is justified by the fact that disproportionately many workers are needed in order to speed up the task execution. Indeed, if this was not the case, the resource allocation problem would (in the absence of per-period consumption quotas) become trivial.

In multi-resource allocation problems, we need to specify how the joint deployment of several resources affects the duration of a project activity. In the following, we assume that activity i's duration, $d_i : \mathbb{R}_+^m \times \mathbb{R}_{++} \mapsto \mathbb{R}_+$, is defined as $d_i(x_i; \omega_i) = \omega_i / \rho_i(x_i)$. Here, $\omega_i > 0$ denotes the *work content* of activity i. The work content is dimensionless and can be interpreted as the level of "difficulty" or "complexity" of performing task i. We denote by $x_i = (x_i^1, \ldots, x_i^m) \in \mathbb{R}_+^m$ the subvector of x that describes the resources spent on activity i. The function $\rho_i : \mathbb{R}_+^m \mapsto \mathbb{R}_{++}$ maps an investment vector x_i to its associated "productivity". The inverse-proportional relation between d_i and ρ_i has intuitive appeal since higher productivities should result in shorter task durations. As we will see in the following, this relation preserves desirable properties of the productivity mapping ρ_i.

We are thus led to the problem of specifying appropriate productivity mappings ρ_i. Natural candidates are production functions from microeconomics: a production function determines the output quantity of a production process (e.g., the lot size of a certain product) as a function of the input factors (e.g., the amount of labor and capital employed). In our case, the output is a productivity, that is, the capacity to carry out work that is related to the completion of a project task. Two classes of production functions are common in microeconomics since they describe resource interactions that are often observed in practice [MCWG95]. Limitational functions describe production processes which combine the input factors in a fixed proportion (e.g., cars consist of four tires and one steering wheel). Substitutional functions, on the other hand, reflect processes where the abundance of some input factors can be used to partially offset the shortage of others (e.g., different types of fertilizer in the cultivation of land).

We define *limitational productivity mappings* as

$$\rho_i^{\mathrm{L}}(x_i) = \delta_i \min \left\{ \psi_i^k x_i^k \, : \, k \in K, \; \psi_i^k > 0 \right\}^{\gamma_i}. \tag{5.1}$$

Here, $\delta_i > 0$ describes the efficiency of the process underlying activity i. We can omit this parameter by scaling ω_i. The vector $\psi_i \in \mathbb{R}_+^m$ characterizes the optimal input factor ratios, that is, the investment weights that lead to zero wastage. The exponent $\gamma_i > 0$ determines the degree of homogeneity: for any scaling parameter $\lambda \geq 0$ we have $d_i(\lambda x_i; \omega_i) = \lambda^{-\gamma_i} d_i(x_i; \omega_i)$. Hence, a λ-fold increase of every input factor leads to a λ^{γ_i}-fold decrease in task duration. Limitational productivity mappings have zero substitution elasticity, that is, it is not possible to substitute one input factor by another. The left part of Fig. 5.1 visualizes this type of productivity mapping. In the context of project scheduling, typical examples of limitational

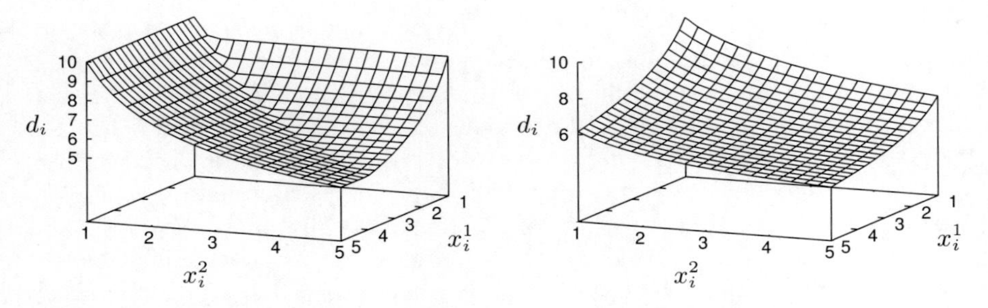

Fig. 5.1 Activity duration depending on two input factors which are combined in a limitational (*left*) and a substitutional (*right*) process. Abundance of a single resource leads to wastage in the former case, whereas it leads to further time savings in the latter one

productivity mappings are predefined team structures (e.g., one foreman and five untrained workers form a team) and the incorporation of machinery or materials (e.g., four workers are required to operate one flexible manufacturing system). One can show that if all activity durations are determined by limitational productivity mappings, then the allocation problem can be reformulated as a single-resource problem.

We define *substitutional (Cobb–Douglas) productivity mappings* as

$$\rho_i^{\mathrm{S}}(x_i) = \delta_i \prod_{k \in K} (x_i^k)^{\psi_i^k}, \tag{5.2}$$

where $\delta_i > 0$ is again an efficiency parameter that can be transformed away. The exponents $\psi_i^k \in \mathbb{R}_+$ specify the partial elasticities of d_i with respect to x_i^k:

$$\frac{\partial d_i(x_i;\omega_i)/\partial x_i^p}{d_i(x_i;\omega_i)/x_i^p} = \frac{-(\omega_i/\delta_i)\psi_i^p (x_i^p)^{-\psi_i^p-1} \prod_{k \neq p} (x_i^k)^{-\psi_i^k}}{(\omega_i/\delta_i)(x_i^p)^{-\psi_i^p-1} \prod_{k \neq p} (x_i^k)^{-\psi_i^k}} = -\psi_i^p.$$

Hence, a marginal increase of x_i^p leads, ceteris paribus, to a ψ_i^p-fold decrease of d_i. We furthermore see that d_i is homogeneous of degree $-\sum_{k \in K} \psi_i^k$; this term has the same interpretation as $-\gamma_i$ in (5.1). The marginal rate of technical substitution (MRTS) of input p for input q amounts to

$$\mathrm{MRTS}_{p,q} = \frac{\partial d_i(x_i;\omega_i)/\partial x_i^p}{\partial d_i(x_i;\omega_i)/\partial x_i^q} = \frac{-(\omega_i/\delta_i)\psi_i^p (x_i^p)^{-\psi_i^p-1} \prod_{k \neq p} (x_i^k)^{-\psi_i^k}}{-(\omega_i/\delta_i)\psi_i^q (x_i^q)^{-\psi_i^q-1} \prod_{k \neq q} (x_i^k)^{-\psi_i^k}} = \frac{\psi_i^p x_i^q}{\psi_i^q x_i^p}.$$

Thus, in order to keep the duration of activity i unchanged, a marginal decrease of x_i^p requires a $(\psi_i^p x_i^q)/(\psi_i^q x_i^p)$-fold increase of x_i^q. The right part of Fig. 5.1 visualizes the Cobb–Douglas productivity mapping. In project scheduling, substitutional productivity mappings arise from outsourcing decisions (part of an activity is done in-house, the rest is outsourced), flexible degrees of automation (labor and capital are often substitutes within certain ranges) and flexible team structures (several untrained workers can replace a trained worker).

In the following, we denote by V^{L} and V^{S} the sets of activities whose durations are determined by limitational and substitutional productivity mappings, respectively. We assume that $V = V^{\mathrm{L}} \cup V^{\mathrm{S}}$ and $V^{\mathrm{L}} \cap V^{\mathrm{S}} = \emptyset$. The resulting deterministic resource allocation model can be described as follows:

$$\underset{x,y}{\text{minimize}} \quad y_n + d_n(x_n; \omega_n) \tag{5.3a}$$

$$\begin{aligned}
\text{subject to} \quad & x \in \mathbb{R}_+^{mn}, \quad y \in \mathbb{R}_+^n \\
& y_j \geq y_i + d_i(x_i; \omega_i) && \forall (i,j) \in E, && (5.3\text{b}) \\
& \sum_{i \in V} x_i^k \leq B_k && \forall k \in K, && (5.3\text{c}) \\
& x_i^k \in \left[\underline{c}_i^k, \overline{c}_i^k\right] && \forall i \in V,\ k \in K. && (5.3\text{d})
\end{aligned}$$

In this model, the decision vector $y \in \mathbb{R}_+^n$ contains the start times of the project activities. The objective is to minimize the project makespan. Since we assume that the temporal network has a unique sink node n, this is equivalent to minimizing the completion time of activity n. Constraint (5.3b) enforces the temporal precedences between the project tasks, while constraints (5.3c) and (5.3d) enforce the budget and process constraints, respectively. For future use, we define $X = \{x \in \mathbb{R}_+^{mn} : x \text{ satisfies (5.3c) and (5.3d)}\}$.

From a computational viewpoint, the following observation is crucial.

Proposition 5.2.1. *Problem (5.3) can be formulated as a convex optimization model.*

Proof. Without loss of generality, we can assume that $\omega_n = 0$. As a result, the only nonlinearity occurs in constraint (5.3b). By a slight abuse of notation, we introduce variables $d \in \mathbb{R}_+^n$ for the task durations and replace constraint (5.3b) with

$$\begin{aligned}
& y_j \geq y_i + d_i && \forall (i,j) \in E, && (5.3\text{b'}) \\
& d_i \rho_i(x_i) \geq \omega_i && \forall i \in V. && (5.3\text{b''})
\end{aligned}$$

Because we are minimizing the project's makespan, there is always an optimal solution to the new model that satisfies (5.3b") as equality. This establishes the equivalence to the original model. By construction, $d_i \rho_i(x_i)$ is log-concave in (d_i, x_i) for substitutional activities. For a limitational activity, we note that

$$d_i \rho_i^{\mathrm{L}}(x_i) \geq \omega_i \quad \Longleftrightarrow \quad d_i \delta_i (\psi_i^k x_i^k)^{\gamma_i} \geq \omega_i \quad \forall k \in K : \psi_i^k > 0,$$

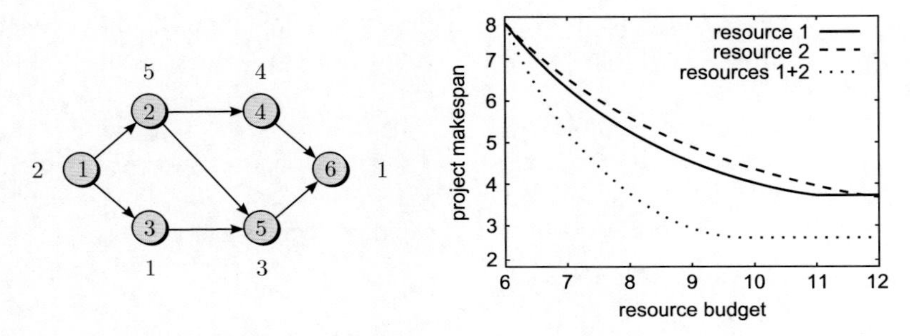

Fig. 5.2 Deterministic resource allocation for an example project. The *left chart* illustrates the project network and the activities' work contents (attached to the nodes). The *right chart* presents the project's makespan as a function of the resource budgets

and the left-hand side of the latter constraint is log-concave in (d_i, x_i) as well. Thus, the feasible region of the extended optimization problem is convex. □

We illustrate model (5.3) with an example.

Example 5.2.1. Consider the project in Fig. 5.2, left. Apart from the missing cash flows, it is identical to the temporal network in Fig. 1.1. Now, however, we interpret the numbers attached to the network tasks as the work contents of the project activities. We consider two resources with process constraints $x^k \in [(1/4)\mathrm{e}, 2\mathrm{e}]$ and budget constraints $\sum_{i \in V} x_i^k \leq 6,\ k \in \{1, 2\}$. Activity 4 has a limitational productivity mapping, whereas all other activities are described by substitutional productivity mappings:

$$\rho_i(x_i) = \begin{cases} \min\left\{2x_i^1,\ x_i^2\right\} & \text{if } i = 4, \\ \left(x_i^1\right)^2 \left(x_i^2\right)^{3/2} & \text{otherwise.} \end{cases}$$

Since the work content attached to the sink node 6 is nonzero, we introduce an auxiliary variable τ that represents the completion of the project. Using the reformulation from Proposition 5.2.1, we can then formulate the deterministic resource allocation model (5.3) as follows:

$$\begin{aligned}
\underset{d,x,y,\tau}{\text{minimize}} \quad & \tau \\
\text{subject to} \quad & d \in \mathbb{R}_+^6, \quad x \in \mathbb{R}_+^{12}, \quad y \in \mathbb{R}_+^6, \quad \tau \in \mathbb{R}_+ \\
& y_2 \geq y_1 + d_1, \quad y_3 \geq y_1 + d_1, \quad y_4 \geq y_2 + d_2, \\
& y_5 \geq y_2 + d_2, \quad y_5 \geq y_3 + d_3, \quad y_6 \geq y_4 + d_4, \\
& y_6 \geq y_5 + d_5, \quad \tau \geq y_6 + d_6, \\
& d_1 \left(x_1^1\right)^2 \left(x_1^2\right)^{3/2} \geq 2, \quad d_2 \left(x_2^1\right)^2 \left(x_2^2\right)^{3/2} \geq 5,
\end{aligned}$$

$$d_3\left(x_3^1\right)^2\left(x_3^2\right)^{3/2} \geq 1, \quad d_5\left(x_5^1\right)^2\left(x_5^2\right)^{3/2} \geq 3,$$
$$d_6\left(x_6^1\right)^2\left(x_6^2\right)^{3/2} \geq 1,$$
$$2d_4x_4^1 \geq 4, \quad d_4x_4^2 \geq 4,$$
$$x^1, x^2 \in [(1/4)\mathrm{e}, 2\mathrm{e}], \quad \sum_{i=1}^{6} x_i^1 \leq 6, \quad \sum_{i=1}^{6} x_i^2 \leq 6.$$

The optimal resource allocation for this problem is

$$x^1 \approx (1.22, 1.38, 0.80, 0.52, 1.03, 1.05)$$
$$\text{and} \quad x^2 \approx (1.10, 1.25, 0.73, 1.05, 0.93, 0.95),$$

and the associated makespan is $\tau \approx 7.85$.

Let us now investigate the impact of the resource budgets on the project's makespan. To this end, two of the curves in Fig. 5.2, right vary the budget of one resource while the other budget is fixed at 6. The third curve simultaneously varies the budget of both resources. Clearly, the project makespan decreases if we increase the resource budgets. If we only increase the budget of resource 1, then resource 2 soon becomes a bottleneck and we cannot decrease the makespan beyond 3.72. This is due to the fact that task 4 requires a larger amount of resource 2. If we simultaneously increase the budget of both resources, however, we can avoid this bottleneck by substituting resource 2 with resource 1 in the activities $i \in V \setminus \{4\}$. We obtain the minimal project makespan 2.71 by assigning a budget of 9.8 to both resources.

We close with some remarks about the deterministic resource allocation model. Firstly, if all productivity mappings contain rational exponents, then model (5.3) can be formulated as a conic quadratic program [AG03]. Depending on the values of these exponents, this can lead to performance improvements over solving the model with a general convex optimizer.

Secondly, model (5.3) only accommodates simple productivity mappings. Sometimes one may require nested productivity mappings that map investment levels and/or productivity values to (new) productivity values. For example, a trade-off between a limitational labor process (e.g., foreworkers and untrained labor have to satisfy a proportion of 1:4) and capital (e.g., an outsourcing decision) can be modeled as a two-stage process. It is easy to extend model (5.3) to nested productivity mappings such that the resulting problem remains convex and representable as a conic quadratic program.

Finally, the parameter values of the productivity mappings might be unavailable in practice. Nevertheless, one can assume that at least the type of productivity mapping (limitational or substitutional) is known for each activity. With this knowledge, one can estimate the missing parameter values based on a set of expected durations for different resource combinations.

5.3 Resource Allocation Under Uncertainty

In the remainder of this chapter, we assume that the vector of work contents constitutes a random vector with finite first and second moments. To emphasize the difference to the deterministic setting in the previous section, we use the notation $\widetilde{\omega}$ for the work contents from now on. By convention, all random objects in this chapter, which are indicated by the tilde sign, are defined on an abstract probability space $(\Omega, \mathcal{F}, \mathbb{P})$. In contrast to the work contents, all other parameters remain deterministic. For references to models in which the project graph G or the process and budget constraints are uncertain, see Sect. 2.3. It seems most appropriate to treat the parameters of the productivity mappings as deterministic quantities. In fact, it is unlikely that the decision maker can specify meaningful distributions for them. Moreover, the impact of uncertain work contents can outweigh by far the consequences of not knowing the exact productivity mappings. Thus, little accuracy may be lost when the latter are assumed to be deterministic.

We consider static resource allocations that are chosen before any of the uncertain work contents is revealed. The corresponding decision vector x is thus a here-and-now decision, see Sect. 2.2.1. Static allocations are frequently required when some resources (such as labor and machinery) cannot be shifted between different activities on short notice [HL05]. Even if it was admissible to adapt the resource allocation during project implementation, a static allocation might still be preferable from the viewpoint of computational tractability, see Sect. 2.2.3 and [GG06, JWW98]. Contrary to the resource allocation x, we assume that the activity start times y are allowed to depend on the realization of the uncertain work contents $\widetilde{\omega}$. In the terminology of Sect. 2.2.1, y is thus a wait-and-see decision. Indeed, if y was modeled as a here-and-now decision, we would seek for a schedule of a priori fixed activity start times that can always (or with high reliability) be met [HL05]. Since we assume the absence of consumption quotas per unit time (see Sect. 5.1), however, there is no benefit in knowing the activity start times before the implementation of the project. Thus, fixed start times would unnecessarily increase the project's makespan in our setting.

In Sect. 2.2.1 we defined the α-VaR of a random variable as the α-quantile of its probability distribution. In the face of uncertainty about the work contents, our new goal is to minimize the α-VaR of the random project makespan. This results in the following reformulation of the deterministic model (5.3).

$$\underset{x,\tau}{\text{minimize}} \quad \tau \tag{5.4a}$$

$$\text{subject to} \quad x \in \mathbb{R}_{+}^{mn}, \quad \tau \in \mathbb{R}_{+}$$

$$\mathbb{P}\left(\exists\, y \geq 0 : \left\{ \begin{array}{ll} \tau \geq y_n + d_n(x_n; \widetilde{\omega}_n) & \\ y_j \geq y_i + d_i(x_i; \widetilde{\omega}_i) & \forall (i,j) \in E \end{array} \right\} \right) \geq \alpha, \tag{5.4b}$$

$$x \in X. \tag{5.4c}$$

Model (5.4) constitutes a two-stage chance constrained stochastic program, see Sect. 2.2.1. The uncertain work contents $\widetilde{\omega}$ are revealed after the resource allocation x has been chosen, but before the activity start times y are selected. The joint chance constraint (5.4b) ensures that τ is a valid upper bound on the project makespan with probability at least α. Since τ is minimized, model (5.4) indeed minimizes the α-VaR of the project makespan.

The fact that y is chosen after *all* uncertain work contents have been revealed seems to violate non-anticipativity [Pré95, RS03]: in order to be implementable, the start time y_j of activity j must only depend on those work contents that are known at the time when j is started. The uncertain work content of an activity, however, is only known after its completion. Since model (5.4) principally allows y_j to depend on all components of $\widetilde{\omega}$, the resulting optimal policy could therefore be acausal. Fortunately, it turns out that similar to our findings in Chap. 4, the non-anticipative early start schedule is always among the optimal solutions to problem (5.4). Since the project graph is assumed to be acyclic, the early start schedule can be calculated recursively via

$$y_j^*(x;\widetilde{\omega}) = \max\left\{0, \sup_{i\in V}\{y_i^*(x;\widetilde{\omega}) + d_i(x_i;\widetilde{\omega}_i) \,:\, (i,j)\in E\}\right\} \quad \forall j\in V$$

for every fixed x. Note that this schedule is non-anticipative since the start time of an activity only depends on the completion times of predecessor activities, that is, only knowledge about the work contents of completed activities is required. Furthermore, the absence of per-period resource consumption quotas guarantees that the early start schedule is always feasible. Finally, since the makespan is a non-decreasing function of the activity start times, the early start schedule minimizes the makespan of the project for any fixed x and any fixed realization of $\widetilde{\omega}$. Hence, if an optimal solution to problem (5.4) contains an anticipative start time schedule y, we can replace it with the (non-anticipative) early start schedule without sacrificing optimality.

Two-stage chance constrained problems of type (5.4) are notoriously difficult to solve [EI07]. Several approximate solution methods have been suggested in the literature, such as sampling-based variants of the ellipsoid method [EI07, NS06b], convex approximation via CVaR constraints [WA08] and methods based on affine decision rules [CSS07]. In the following, we will consider a reformulation of problem (5.4) that eliminates the two-stage structure. We will compare this approach with direct approximations of (5.4) via CVaR constraints in Sect. 5.4. Affine decision rules are studied in Chap. 6.

We eliminate the two-stage structure of problem (5.4) by enumerating the activity paths of the project graph. Apart from reducing the model to a single-stage problem, this approach enables us to employ normal approximations for the distributions of the path durations that can be justified by a generalized central limit theorem. The number of activity paths can be exponential in the size of the project graph, see Chap. 6. Since the considered reformulation will contain one constraint per activity path, this implies that the model can potentially contain an exponential number of constraints. Sometimes, however, project instances may contain only moderate numbers of activity paths. Furthermore, we will discuss a technique which alleviates the problem of large path numbers in Sect. 5.5.2. We caution the reader that in other application areas of temporal networks, the number of network paths can be huge. In Chap. 6 we will discuss a technique to minimize the worst-case makespan in networks with large numbers of paths.

We recall that a path in a directed graph $G = (V, E)$ constitutes a list of nodes $(i_1, \ldots, i_p)$ such that $(i_1, i_2), \ldots, (i_{p-1}, i_p) \in E$. We define an activity path $P = \{i_1, \ldots, i_p\} \subseteq V$ as a set of project activities that form a path in the project graph G. We denote by $\overline{\mathcal{P}} = \{P^l\}_l$ the set of inclusion-maximal paths, that is, $\overline{\mathcal{P}}$ contains all activity paths that are not strictly included in any other path. Observe that for a fixed vector of work contents, a project's makespan equals the duration of the most time-consuming path for those work contents. Hence, we can reformulate problem (5.4) equivalently as follows:

$$\underset{x,\tau}{\text{minimize}} \quad \tau \tag{5.5a}$$

$$\text{subject to} \quad x \in \mathbb{R}^{mn}_{+}, \quad \tau \in \mathbb{R}_{+}$$

$$\mathbb{P}\Big(\tau \geq \sum_{i \in P^l} d_i(x_i; \widetilde{\omega}_i) \quad \forall P^l \in \overline{\mathcal{P}}\Big) \geq \alpha, \tag{5.5b}$$

$$x \in X. \tag{5.5c}$$

Note that problem (5.5) only involves here-and-now decisions (x, τ) and hence constitutes a single-stage chance constrained problem. Similar to constraint (5.4b), however, the constraint (5.5b) still constitutes a joint chance constraint in which the random variables cannot easily be separated from the decision variables. Apart from some benign special cases, problems of type (5.5) generically have nonconvex or even disconnected feasible sets, which severely complicates their numerical solution. Well-structured chance constrained problems that have convex feasible sets for all or for sufficiently high values of α are discussed in [HS08, Pré95]. One readily verifies, however, that model (5.5) does not belong to these problem classes. The following example shows that model (5.5) is indeed nonconvex.

Example 5.3.1. Consider the project $G = (V, E)$ with node set $V = \{1, \ldots, 4\}$ and precedences $E = \{(1,2), (1,3), (2,4), (3,4)\}$. For the sake of simplicity, let us assume that $\widetilde{\omega}_1 = \widetilde{\omega}_4 = 0$ almost surely, $\widetilde{\omega}_2$ and $\widetilde{\omega}_3$ follow independent standard normal distributions and $\rho_i(x_i) = x_i$, $i = 2, 3$. The process constraints are $1/2 \leq x_2, x_3 \leq 2$, and there is no resource budget. We want to investigate the convexity of the feasible region

$$X(\alpha) = \Big\{(\tau, x_2, x_3) \in \mathbb{R}_+ \times [1/2, 2]^2 \,:\, \mathbb{P}\left(\tau \geq \widetilde{\omega}_2/x_2,\ \tau \geq \widetilde{\omega}_3/x_3\right) \geq \alpha\Big\}$$

$$= \Big\{(\tau, x_2, x_3) \in \mathbb{R}_+ \times [1/2, 2]^2 \,:\, \Phi(\tau x_2)\, \Phi(\tau x_3) \geq \alpha\Big\}.$$

It is easy to see that $X(\alpha)$ is generically nonconvex. Indeed, for $\alpha = 2/3$ one can verify that $(\tau^1, x_2^1, x_3^1) = (1, 2, 1/2), (\tau^2, x_2^2, x_3^2) = (1, 1/2, 2) \in X(2/3)$, but $(\tau, x_2, x_3) = 1/2(\tau^1, x_2^1, x_3^1) + 1/2(\tau^2, x_2^2, x_3^2) = (1, 5/4, 5/4) \notin X(2/3)$.

Recently, sample approximation [LA08] and scenario approximation techniques [CC05, CC06] have been proposed for solving joint chance constrained problems of type (5.5). Applied to our setting, however, sample approximation would lead to large mixed-integer nonlinear programs (even in the absence of discrete resources), which themselves constitute difficult optimization problems. Likewise, solving model (5.5) with scenario approximation techniques would result in a problem whose number of constraints is proportional to the cardinality of $\overline{\mathcal{P}}$ times the number of scenarios employed. Since this product is large in realistic settings, this approach seems primarily interesting for small projects.

In this chapter, we employ Boole's inequality to approximate problem (5.5) as follows:

$$\underset{x,\beta,\tau}{\text{minimize}} \quad \tau \tag{5.6a}$$

$$\text{subject to} \quad x \in \mathbb{R}_+^{mn}, \quad \beta \in \mathbb{R}_+^{|\overline{\mathcal{P}}|}, \quad \tau \in \mathbb{R}_+$$

$$\mathbb{P}\Big(\tau \geq \sum_{i \in P^l} d_i(x_i; \widetilde{\omega}_i)\Big) \geq \beta_l \qquad \forall P^l \in \overline{\mathcal{P}}, \tag{5.6b}$$

$$\sum_{P^l \in \overline{\mathcal{P}}} \beta_l \geq \alpha + (|\overline{\mathcal{P}}| - 1), \tag{5.6c}$$

$$\beta_l \in [0, 1] \qquad \forall P^l \in \overline{\mathcal{P}}, \tag{5.6d}$$

$$x \in X. \tag{5.6e}$$

For future use, we define $B = \{\beta \in \mathbb{R}_+^{|\overline{\mathcal{P}}|} : \beta \text{ satisfies (5.6c) and (5.6d)}\}$. Note that in the constraint (5.6b) we have split up the joint chance constraint of model (5.5) into independent separated chance constraints. The following proposition shows that model (5.6) constitutes a conservative approximation of problem (5.5). The proof follows the arguments presented in [NS06a].

Proposition 5.3.1. *If (x, β, τ) is a feasible solution to model (5.6), then (x, τ) is also feasible in model (5.5).*

Proof. Using the feasibility of (x, β, τ) in problem (5.6), we find that

$$\mathbb{P}\left(\tau \geq \sum_{i \in P^l} d_i(x_i; \widetilde{\omega}_i) \ \forall P^l \in \overline{\mathcal{P}}\right) = 1 - \mathbb{P}\left(\bigcup_{P^l \in \overline{\mathcal{P}}} \left\{\tau < \sum_{i \in P^l} d_i(x_i; \widetilde{\omega}_i)\right\}\right)$$

$$\geq 1 - \sum_{P^l \in \overline{\mathcal{P}}} \mathbb{P}\left(\tau < \sum_{i \in P^l} d_i(x_i;\widetilde{\omega}_i)\right) = 1 - \sum_{P^l \in \overline{\mathcal{P}}} \left[1 - \mathbb{P}\left(\tau \geq \sum_{i \in P^l} d_i(x_i;\widetilde{\omega}_i)\right)\right]$$

$$\geq 1 - \sum_{P^l \in \overline{\mathcal{P}}} (1 - \beta_l) = 1 - |\overline{\mathcal{P}}| + \sum_{P^l \in \overline{\mathcal{P}}} \beta_l \ \geq \alpha.$$

Here, the first inequality follows from Boole's inequality.[1] □

Observe that for $\alpha < 1$, both of the problems (5.5) and (5.6) are feasible if and only if $X \neq \emptyset$. However, the optimal objective value of problem (5.6) is greater than or equal to the optimal objective value of problem (5.5). The approximation (5.6) can in principle be tightened by incorporating pairs of activity paths via Bonferroni's inequalities, see [Pré95]. This, however, either requires an a priori fixed choice of admissible path pairs or a selection procedure that determines optimal pairs in an iterative manner. The former approach is likely to result in a substantial increase of problem size, while the latter technique requires the repeated solution of model (5.6). Since Boole's approximation turns out to be remarkably tight in our numerical tests (see Sect. 5.4), the potential gains of either approach are likely to be outweighed by the increase in complexity. Hence, we settle for Boole's inequality in the following.

Model (5.6) still constitutes a generically nonconvex problem. More so, even the verification whether a given point is feasible requires the evaluation of multi-dimensional integrals and thus becomes prohibitively expensive for realistic problem sizes. In recent years, several inequalities from probability theory have been employed to obtain conservative convex approximations of separated chance constraints [CSS07, NS06a]. In the following, we will pursue a different approach, namely, we will simplify constraint (5.6b) by approximating the path durations $\sum_{i \in P^l} d_i(x_i;\widetilde{\omega}_i)$, $P^l \in \overline{\mathcal{P}}$, via normal distributions. As we will see, this approximation has theoretical appeal and leads to superior results in numerical tests.

Let the first and second moments of $\widetilde{\omega}$ be given by $\mu = (\mathbb{E}[\widetilde{\omega}_1], \ldots, \mathbb{E}[\widetilde{\omega}_n])^\top$ and $\Sigma \in \mathbb{R}^{n \times n}$, $\Sigma_{ij} = \operatorname{Cov}(\widetilde{\omega}_i, \widetilde{\omega}_j)$. In order to simplify the notation, we furthermore introduce functions $\varrho_l : \mathbb{R}^{mn}_{+} \mapsto \mathbb{R}^n_{+}$, $P^l \in \overline{\mathcal{P}}$, with

$$[\varrho_l(x)]_i = \begin{cases} 1/\rho_i(x_i) & \text{if } i \in P^l, \\ 0 & \text{otherwise.} \end{cases}$$

Using this notation, we can express the mean and the variance of the path duration $\sum_{i \in P^l} d_i(x_i;\widetilde{\omega}_i)$ as $\mu^\top \varrho_l(x)$ and $\varrho_l(x)^\top \Sigma \varrho_l(x)$, respectively, for each $P^l \in \overline{\mathcal{P}}$. In the remainder of this section, we will present a solution method for problem (5.6) that approximates the duration of path P^l by a normal distribution with the same first and second moments. The following generalized central limit theorem justifies

[1] Boole's inequality: For a countable set of events $A_1, A_2, \ldots \in \mathcal{F}$, $\mathbb{P}\left(\bigcup_i A_i\right) \leq \sum_i \mathbb{P}(A_i)$.

the use of such normal approximations in project scheduling under three alternative regularity conditions.

Theorem 5.3.1. *Let* $P_\nu = \{1, \ldots, \nu\}$, $\nu = 1, 2, \ldots$, *be an inclusion-increasing sequence of project paths with task durations* $d_i(x_i; \widetilde{\omega}_i) = \widetilde{\omega}_i / \rho_i$, $\rho_i \in \left[\underline{\rho}, \overline{\rho}\right]$, $\underline{\rho} > 0$, *and* $\widetilde{\omega}_i \geq \underline{\omega} > 0$ $\mathbb{P}$-*a.s. for all* i. *Assume that the first three moments of* $\widetilde{\omega}_i$ *are finite and satisfy*

$$\begin{aligned}
\mu_i &= \mathbb{E}(\widetilde{\omega}_i) \leq \overline{\mu}, \\
\sigma_i^2 &= \mathrm{Var}(\widetilde{\omega}_i) \in [\underline{\sigma}^2, \overline{\sigma}^2] \quad \textit{with } \underline{\sigma}^2 > 0 \\
\textit{and} \quad \gamma_i^3 &= \mathbb{E}\left(|\widetilde{\omega}_i - \mu_i|^3 \right) \leq \overline{\gamma}^3.
\end{aligned}$$

Then for any fixed resource allocation, the standardized path durations converge in distribution to a standard normal distribution as $\nu \longrightarrow \infty$ *if either of the following three conditions holds:*

(C1) The components of $\widetilde{\omega}$ *follow a multivariate normal distribution.*
(C2) The components of $\widetilde{\omega}$ *are independent.*
(C3) There is a time lag $T \in \mathbb{R}_+$ *such that* $\widetilde{\omega}_i$ *and* $\widetilde{\omega}_j$ *are independent if the start times of tasks* i *and* j *differ by at least* T *time units. Furthermore, the covariances of dependent work contents are bounded from above by some* $\zeta \in \mathbb{R}_+$, *and* $\lim_{\nu \longrightarrow \infty} \nu^{-1} \mathrm{Var}(\sum_{i \in P_\nu} d_i(x_i; \widetilde{\omega}_i))$ *exists and is nonzero.*

Proof. Since the duration of any project path is linear in $\widetilde{\omega}$, it is normally distributed if $\widetilde{\omega}$ follows a multivariate normal distribution. Thus, we obtain the stronger result that under (C1), all path durations are normally distributed.

In the following, we abbreviate $d_i(x_i; \widetilde{\omega}_i)$ by $\widetilde{d}_i$. Under assumption (C2), the assertion follows from Lyapunov's central limit theorem [Pet75]. Apart from finite first and second moments (which are implied by our assumptions), this theorem requires that

$$\lim_{\nu \longrightarrow \infty} \frac{\left(\sum_{i \in P_\nu} \mathbb{E}\left(|\widetilde{d}_i - \mathbb{E}(\widetilde{d}_i)|^3 \right) \right)^{1/3}}{\left(\sum_{i \in P_\nu} \mathrm{Var}\left(\widetilde{d}_i \right) \right)^{1/2}} = 0.$$

Employing the estimates

$$\mathbb{E}\left(|\widetilde{d}_i - \mathbb{E}(\widetilde{d}_i)|^3 \right) \leq \frac{1}{\underline{\rho}^3} \mathbb{E}\left(|\widetilde{\omega}_i - \mathbb{E}(\widetilde{\omega}_i)|^3 \right) \leq \frac{\overline{\gamma}^3}{\underline{\rho}^3}$$

and

$$\mathrm{Var}(\widetilde{d}_i) \geq \frac{1}{\overline{\rho}^2} \mathrm{Var}(\widetilde{\omega}_i) \geq \frac{\underline{\sigma}^2}{\overline{\rho}^2},$$

we obtain

$$\frac{\left(\sum_{i\in P_\nu}\mathbb{E}\left(\left|\widetilde{d}_i-\mathbb{E}(\widetilde{d}_i)\right|^3\right)\right)^{1/3}}{\left(\sum_{i\in P_\nu}\mathrm{Var}(\widetilde{d}_i)\right)^{1/2}}\leq\frac{\left(\nu\overline{\gamma}^3/\underline{\rho}^3\right)^{1/3}}{\left(\nu\underline{\sigma}^2/\overline{\rho}^2\right)^{1/2}}=\nu^{-\frac{1}{6}}\frac{\overline{\gamma}\,\overline{\rho}}{\underline{\sigma}\,\underline{\rho}},$$

and the last term indeed converges to zero for $\nu\longrightarrow\infty$.

Under condition (C3), the claim follows from Berk's central limit theorem for m-dependent random variables, see [Ber73]. Translated into our context, this theorem is based on the following assumptions:

(A1) There is $m\in\mathbb{N}_0$ such that $\widetilde{d}_i$ and $\widetilde{d}_j$ are independent if $|i-j|>m$.
(A2) $\mathbb{E}(|\widetilde{d}_i|^3)$ is uniformly bounded for all i.
(A3) $\mathrm{Var}(\widetilde{d}_{i+1}+\ldots+\widetilde{d}_j)\leq(j-i)M$ for some $M\in\mathbb{R}_+$ and all i,j.
(A4) $\lim_{\nu\longrightarrow\infty}\nu^{-1}\mathrm{Var}(\sum_{i\in P_\nu}\widetilde{d}_i)$ exists and is nonzero.

By condition (C3), $\widetilde{\omega}_i$ and $\widetilde{\omega}_j$ are independent if the start times of the respective activities differ by at least T time units. For $i<j$, the start time difference between activities i and j amounts to at least $\sum_{l=i}^{j-1}\widetilde{d}_l\geq(j-i)\underline{\omega}/\overline{\rho}$. Thus, $m=\lceil(T\overline{\rho})/\underline{\omega}\rceil$ is sufficient to guarantee independence of $\widetilde{\omega}_i$ and $\widetilde{\omega}_j$ (and hence, of $\widetilde{d}_i$ and $\widetilde{d}_j$) whenever $|i-j|>m$, as required by (A1). Concerning (A2), we see that

$$0\leq\mathbb{E}\left(\left|\widetilde{d}_i\right|^3\right)\leq\frac{1}{\underline{\rho}^3}\mathbb{E}\left(\widetilde{\omega}_i^3\right).$$

Since $\mathbb{E}(|\widetilde{\omega}_i-\mu_i|^3)$ is uniformly bounded for all i, so is $\mathbb{E}(\widetilde{\omega}_i^3)$. As for (A3), we note that

$$\begin{aligned}\mathrm{Var}\left(\widetilde{d}_{i+1}+\ldots+\widetilde{d}_j\right)&=\sum_{p=i+1}^{j}\left(\frac{1}{\rho_p^2}\mathrm{Var}(\widetilde{\omega}_p)+2\sum_{q=p+1}^{\min\{j,p+m\}}\frac{1}{\rho_p\rho_q}\mathrm{Cov}(\widetilde{\omega}_p,\widetilde{\omega}_q)\right)\\&\leq\frac{1}{\underline{\rho}^2}\sum_{p=i+1}^{j}\left(\mathrm{Var}(\widetilde{\omega}_p)+2\sum_{q=p+1}^{\min\{j,p+m\}}\mathrm{Cov}(\widetilde{\omega}_p,\widetilde{\omega}_q)\right)\\&\leq\frac{1}{\underline{\rho}^2}\sum_{p=i+1}^{j}(\overline{\sigma}^2+2m\zeta)\leq(j-i)M\end{aligned}$$

for $M=(\overline{\sigma}^2+2m\zeta)/\underline{\rho}^2$. Finally, (A4) directly follows from (C3). □

Condition (C3) is particularly appealing for project scheduling since typical sources of uncertainty (such as weather conditions, staff holidays and illness) tend to be of temporary nature. Apart from the requirement that the limit of $\nu^{-1}\mathrm{Var}(\sum_{i\in P_\nu}d_i(x_i;\widetilde{\omega}_i))$ for $\nu\longrightarrow\infty$ exists, the assumptions of (C3) are rather mild and do not require further explanation. Note that the aforementioned limit is likely to exist in all but pathological cases. It exists, for example, when the (co-)variances of dependent work contents can themselves be regarded as random variables with distributions that satisfy the assumptions of a central limit theorem. However, the limit does not exist if the task durations are independent random variables with variances

$$\mathrm{Var}\big(d_i(x_i;\widetilde{\omega}_i)\big) = \begin{cases} a & \text{if } i \in [2^{2k}, 2^{2k+1}) \text{ for some } k \in \mathbb{N}_0 \\ b & \text{if } i \in [2^{2k+1}, 2^{2k+2}) \text{ for some } k \in \mathbb{N}_0 \end{cases} \quad \text{with } a < b.$$

Indeed, one can show that in this case

$$\frac{1}{\nu}\mathrm{Var}\Big(\sum_{i \in P_\nu} d_i(x_i;\widetilde{\omega}_i)\Big) \begin{cases} \leq \frac{5}{8}a + \frac{3}{8}b & \text{if } \nu = 2^{2k+1} - 1 \text{ for some } k \in \mathbb{N}_0, \\ \geq \frac{3}{8}a + \frac{5}{8}b & \text{if } \nu = 2^{2k+2} - 1 \text{ for some } k \in \mathbb{N}_0. \end{cases}$$

Due to the challenges involved in solving chance constrained problems directly, separated chance constraints are frequently approximated by conservative convex constraints that can be expressed in closed form, that is, without sampling. Such approximations are based on inequalities from probability theory [CSS07, NS06a]. In the following, we compare the quality of several such approximations with the normal approximation presented in this chapter. We consider a project path with five activities and a fixed resource allocation. Figure 5.3 illustrates the probability density functions of the activity durations. Note that we deliberately chose distributions that significantly deviate from normal distributions. Furthermore, a path with five activities is very short and hence seemingly unsuited for normal approximation. Figure 5.4 compares the error of several popular approximations with the normal approximation for both independent and dependent activity durations. Unlike these approximations, the normal approximation does not provide a conservative estimate of the path duration. However, it approximates the true cumulative distribution function significantly better than all other approximations considered. This might be surprising since normal approximations cannot be expected to correctly predict the tail probabilities of generic random variables. The reason for the high accuracy observed here is that project path durations are composite random variables whose components (i.e., the activity durations) are typically of the same order of magnitude and follow smooth, close-to-unimodal distributions. Furthermore, although activity durations may exhibit interdependencies, durations of tasks that are well separated in time can essentially be regarded as independent. We remark that for the probabilities of interest (i.e., $\alpha \geq 0.9$), the only reasonably tight bound is obtained by Chernoff's inequality. This inequality, however, requires complete knowledge about the entire moment generating function of the path duration. Compared to this, the normal approximation poses a very modest burden to the decision maker by requiring information about the first two moments of the work contents. In comparison, Chebychev's (single-sided) inequality assumes knowledge about the first two moments, Markov's inequality about the first moments and Hoeffding's inequality about the first moments and the supports of the activity durations. Note that Hoeffding's inequality requires independence among the activity durations.

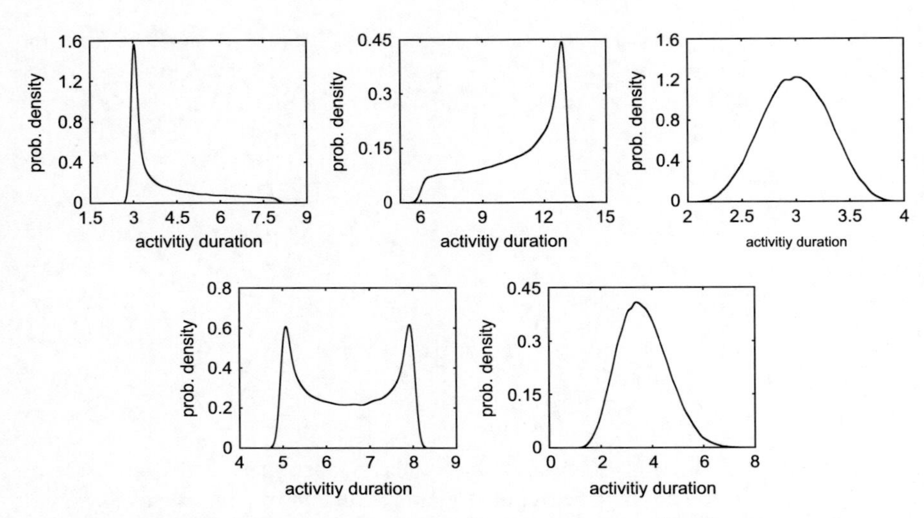

Fig. 5.3 Probability density functions for five activity durations $\{\widetilde{d}_i\}_{i=1}^5$. In the subsequent comparison, we use these durations both directly ("independent" durations) and as disturbances in a first-order autoregressive process $\{\widetilde{d}'_i\}_{i=1}^5$ ("dependent" durations) where $\widetilde{d}'_1 = \widetilde{d}_1$ and $\widetilde{d}'_i = 1/3\widetilde{d}'_{i-1} + 2/3\widetilde{d}_i, i = 2, \ldots, 5$

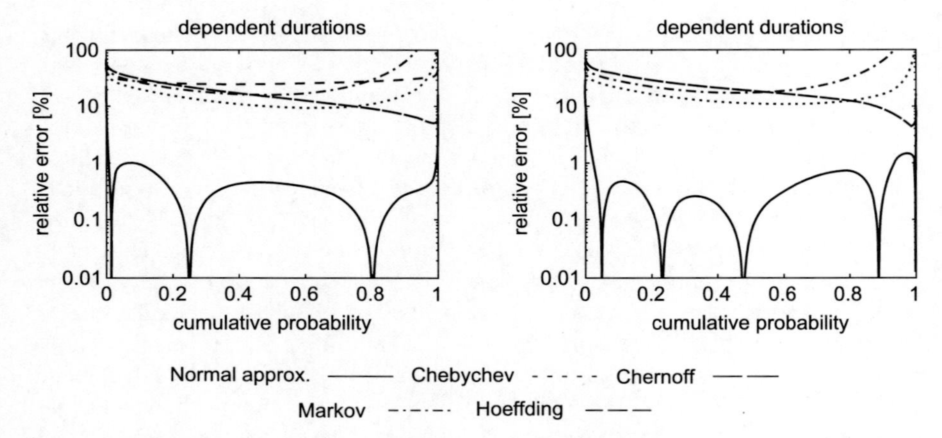

Fig. 5.4 Approximation error of inequalities from probability theory and the normal approximation for independent (*left*) and dependent (*right*) activity durations

Summing up, our preliminary conclusion (which will be supported by the numerical results in Sect. 5.4) is that normal approximations seem well suited to simplify the chance constraints appearing in (5.6b).

Under the normal approximation, the individual (path-wise) chance constraints in (5.6b) are replaced with

$$\Phi\left(\frac{\tau - \mu^\top \varrho_l(x)}{\sqrt{\varrho_l(x)^\top \, \Sigma \, \varrho_l(x)}}\right) \geq \beta_l \iff \tau \geq \mu^\top \varrho_l(x) + \Phi^{-1}(\beta_l) \sqrt{\varrho_l(x)^\top \, \Sigma \, \varrho_l(x)},$$

where Φ denotes the cumulative distribution function of the standard normal distribution. This leads to the following approximation of problem (5.6):

$$\underset{x,\beta,\tau}{\text{minimize}} \quad \tau \tag{5.7a}$$

$$\text{subject to} \quad x \in \mathbb{R}^{mn}_{+}, \quad \beta \in \mathbb{R}^{|\mathcal{P}|}_{+}, \quad \tau \in \mathbb{R}_{+}$$

$$\tau \geq \mu^{\top} \varrho_l(x) + \Phi^{-1}(\beta_l) \sqrt{\varrho_l(x)^{\top} \Sigma \, \varrho_l(x)} \qquad \forall P^l \in \mathcal{P}, \tag{5.7b}$$

$$x \in X, \quad \beta \in B. \tag{5.7c}$$

The following example shows that model (5.7) is still generically nonconvex.

Example 5.3.2. Consider again the project $G = (V, E)$ from Example 5.3.1, that is, $V = \{1, \ldots, 4\}$, $E = \{(1,2), (1,3), (2,4), (3,4)\}$, $\widetilde{\omega}_1 = \widetilde{\omega}_4 = 0$ almost surely, and $\widetilde{\omega}_2$ and $\widetilde{\omega}_3$ follow independent standard normal distributions. One readily verifies that the set

$$Y(\alpha) = \left\{ (\tau, x_2, x_3) \in \mathbb{R}_+ \times [1/2, 2]^2 \, : \, \Phi(\tau x_2) + \Phi(\tau x_3) \geq \alpha + 1 \right\}$$

is generically nonconvex. Indeed, for $\alpha = 2/3$, $(\tau^1, x_2^1, x_3^1) = (1, 2, 1/2)$ and $(\tau^2, x_2^2, x_3^2) = (1, 1/2, 2)$ are elements of $Y(2/3)$, but their convex combination $(\tau, x_2, x_3) = 1/2(\tau^1, x_2^1, x_3^1) + 1/2(\tau^2, x_2^2, x_3^2) = (1, 5/4, 5/4)$ is not part of $Y(2/3)$. However, the set $Y(\alpha)$ represents the projection of

$$Z(\alpha) = \left\{ (\tau, x, \beta) \in \mathbb{R}_+ \times [1/2, 2]^4 \times B \, : \, \tau \geq \Phi^{-1}(\beta_1)/x_2, \ \tau \geq \Phi^{-1}(\beta_2)/x_3 \right\}$$

onto (τ, x_2, x_3), and $Z(\alpha)$ equals the feasible region of problem (5.7) for this example. We therefore conclude that problem (5.7) is generically nonconvex.

The next proposition further analyzes the convexity of model (5.7).

Proposition 5.3.2 *Assume that* $\alpha \geq 1/2$ *and* $\Sigma \geq 0$ *component-wise, and let* $(\widehat{x}, \widehat{\beta}, \widehat{\tau})$ *be feasible in problem (5.7). With the additional constraints* $x = \widehat{x}$ *or* $\beta = \widehat{\beta}$*, problem (5.7) becomes convex in* (β, τ) *or* (x, τ)*, respectively.*

Proof. First note that for $\alpha \geq 1/2$, the requirement $\beta \in B$ implies that $\beta_l \geq 1/2$, $P^l \in \mathcal{P}$, in every feasible solution (x, β, τ). In the following, we will exploit the fact that Φ^{-1} is nonnegative and convex on the interval $[1/2, 1]$.

We only need to investigate constraint (5.7b) since the other constraints and the objective function are clearly convex in (x, β, τ). For $x = \widehat{x}$ fixed, convexity of constraint (5.7b) in τ and β follows from the convexity of Φ^{-1} for $\beta_l \geq 1/2$. For $\beta = \widehat{\beta}$ fixed, on the other hand, we introduce auxiliary variables $y \in \mathbb{R}^n_+$ and $z \in \mathbb{R}^{n^2}_+$, as well as auxiliary constraints

$$y_i\,\rho(x_i) \geq \mu_i \quad \forall i \in V \quad \text{and} \quad z_{ij}^2\,\rho_i(x_i)\,\rho_j(x_j) \geq \sigma_{ij} \quad \forall i, j \in V. \tag{5.7e}$$

Similar arguments as in Proposition 5.2.1 can be used to prove the convexity of the constraints (5.7e). Note that the right-hand sides of these constraints are nonnegative and hence, there are always variables y and z that satisfy (5.7e) as equalities. We replace the constraint (5.7b) with

$$\tau \geq \sum_{i \in V} y_i + \Phi^{-1}(\widehat{\beta}_l)\sqrt{\sum_{i,j \in P^l} z_{ij}^2} \qquad \forall P^l \in \overline{\mathcal{P}}. \tag{5.7f}$$

Since the right-hand sides of (5.7f) are non-decreasing in y and z and we minimize the maximum of these right-hand sides, there is always an optimal solution to problem (5.7) that satisfies (5.7e) as equalities. Hence, the constraints (5.7e) and (5.7f) are indeed an equivalent reformulation of the constraint (5.7b). The first term on the right-hand side of (5.7f) is linear, while the second one is the product of a nonnegative scalar with the Frobenius norm of the matrix (z_{ij}). Both terms are manifestly convex. □

Since the activity durations are nonlinear functions of the decision variables, we need to require that the components of Σ are nonnegative in order to guarantee convexity of problem (5.7) in (x, τ) for fixed β. Hence, we have to assume that the work contents of different activities have nonnegative covariances, that is, all activity durations are either independent or positively correlated. This is not a very restrictive assumption when considering typical sources of uncertainty, such as motivational factors, staff availability, weather conditions and interactions between concurrent projects. It is rather unlikely that such a phenomenon increases the difficulty of some tasks but decreases the complexity of other tasks. In fact, it is a standard assumption in the literature that activity durations are independent [CSS07, DH02, HL05], which is a special case of the nonnegativity assumption. An inspection of Proposition 5.3.2 reveals that if $\Sigma \not\geq 0$, then replacing Σ with $\Sigma^+ = \left([\sigma_{ij}]^+\right)$, $[\sigma_{ij}]^+ = \max\{0, \sigma_{ij}\}$, results in a conservative approximation of problem (5.7). Hence, even if the assumption of nonnegative correlations is violated, we can readily construct a reasonable surrogate problem that satisfies this assumption.

Proposition 5.3.2 suggests a sequential convex optimization scheme which optimizes over (x, τ) and (β, τ) in turns, keeping either β or x fixed to the optimal value of the previous iteration. Algorithm 2 provides an outline of such a procedure.

Algorithm 2 is in the spirit of alternate convex search procedures. In the following, we discuss the main properties of this algorithm. For a more detailed study of alternate convex search procedures, see [KPK07]. We say that a feasible solution (x^*, β^*, τ^*) to (5.7) is a *partial optimum* if (x^*, τ^*) minimizes problem (5.7) for $\beta = \beta^*$ fixed and (β^*, τ^*) minimizes problem (5.7) for $x = x^*$ fixed. The following observation shows that partial optimality is a necessary (but not sufficient) condition for local optimality.

Algorithm 2 Sequential convex optimization procedure for model (5.7).

1. **Initialization.** If $\underline{c} \notin X$, then abort: Problem (5.7) is infeasible. Otherwise, set $x^0 = \underline{c}$, $\tau^0 = \infty$ (current objective value) and $t = 1$ (iteration counter).
2. **Optimization over $(\boldsymbol{\beta}, \boldsymbol{\tau})$.** Solve problem (5.7) in (β, τ) with $x = x^{t-1}$ fixed. If the optimal solution (β^*, τ^*) satisfies $\tau^* < \tau^{t-1}$, then set $\beta^t = \beta^*$, otherwise keep $\beta^t = \beta^{t-1}$.
3. **Optimization over $(\boldsymbol{x}, \boldsymbol{\tau})$.** Solve problem (5.7) in (x, τ) with $\beta = \beta^t$ fixed. If the optimal solution (x^*, τ^*) satisfies $\tau^* < \tau^{t-1}$, then set $x^t = x^*$, otherwise keep $x^t = x^{t-1}$. Set $\tau^t = \tau^*$.
4. **Termination.** If $(x^t, \beta^t) = (x^{t-1}, \beta^{t-1})$, then stop. Otherwise, set $t = t + 1$ and go back to Step 2.

Observation 5.3.1 *For $\Sigma \geq 0$ component-wise, a local optimum (x^*, β^*, τ^*) of model (5.7) is a partial optimum.* □

Proof. Let (x^*, β^*, τ^*) be a local optimum. Then (x^*, τ^*) is a local optimum for $\beta = \beta^*$ fixed. Due to Proposition 5.3.2, (x^*, τ^*) is also a global minimizer of (5.7) for $\beta = \beta^*$ fixed. The same reasoning applies to (β^*, τ^*) if we fix x to x^*. Hence, (x^*, β^*, τ^*) satisfies the definition of a partial optimum. □

However, a partial optimum need not be locally optimal even for convex problems, see [KPK07]. The following proposition summarizes the key properties of Algorithm 2.

Proposition 5.3.3 *Algorithm 2 identifies the (in-)feasibility of an instance of problem (5.7) in Step 1. For feasible instances, the following properties are satisfied:*

(P1) A different feasible solution is identified in every (but the last) iteration.
(P2) The objective values $\{\tau^t\}_t$ are monotonically decreasing and convergent.
(P3) If the algorithm terminates in finite time, then the final iterate is a partial optimum of problem (5.7). If the algorithm does not terminate, then every accumulation point of $\{(x^t, \beta^t, \tau^t)\}_t$ is a partial optimum of problem (5.7). Furthermore, all accumulation points have the same objective value.

Proof. If $\underline{c} \in X$, then (x^0, β^0, τ^0) defined through $x^0 = \underline{c}$, $\beta_l^0 = 1 - (1 - \alpha)/\left|\overline{\mathcal{P}}\right|$ for $P^l \in \overline{\mathcal{P}}$ and

$$\tau^0 = \max_{P^l \in \overline{\mathcal{P}}} \left\{ \mu^\top \varrho_l(x^0) + \Phi^{-1}(\beta_l^0) \sqrt{\varrho_l(x^0)^\top \Sigma \, \varrho_l(x^0)} \right\}$$

constitutes a feasible solution to problem (5.7). If $\underline{c} \notin X$, on the other hand, then $X = \emptyset$. Thus, problem (5.7) is feasible if and only if $\underline{c} \in X$, and hence the algorithm correctly identifies the (in-)feasibility of a problem instance in Step 1. Furthermore, the algorithm determines a feasible solution in every iteration since $\tau^t = \tau^{t-1}$ together with $\beta^t = \beta^{t-1}$ and $x^t = x^{t-1}$ are feasible for $x^t = x^{t-1}$ and $\beta^t = \beta^{t-1}$ fixed, respectively.

Since the algorithm stops in Step 4 once $\tau^t \geq \tau^{t-1}$, $\{\tau^t\}_t$ is strictly monotonically decreasing until the penultimate iteration. Together with the feasibility of (x^t, β^t, τ^t) for all t, this proves (P1). Since the sequence $\{\tau^t\}_t$ is also bounded from below (for example by zero), assertion (P2) follows. If the algorithm terminates after finitely many iterations, then (P3) is satisfied by construction. Assume that the algorithm does not terminate. One can show that the algorithmic map of the procedure is closed [BSS06, KPK07]. This implies that $(x^{t+1}, \beta^{t+1}, \tau^{t+1})$ satisfies the termination criterion in Step 4 if we set (x^t, β^t, τ^t) to any accumulation point $(\widehat{x}, \widehat{\beta}, \widehat{\tau})$ of the sequence $\{(x^t, \beta^t, \tau^t)\}_t$. Hence, $(\widehat{x}, \widehat{\beta}, \widehat{\tau})$ satisfies that $(\widehat{x}, \widehat{\tau})$ is a minimizer of problem (5.7) for $\beta = \widehat{\beta}$ fixed and $(\widehat{\beta}, \widehat{\tau})$ is a minimizer of problem (5.7) for $x = \widehat{x}$ fixed. This, however, is just the definition of a partial optimum. □

For a given instance of problem (5.7) one can easily find finite a priori bounds on the problem variables that do not change the set of optimal solutions. In this case, the feasible set of problem (5.7) is compact and the constructed solution sequence contains accumulation points if Algorithm 2 does not terminate. We emphasize again that partial optima need not constitute local optima of problem (5.7). Note, however, that even the verification whether a particular solution to a biconvex problem is locally optimal is $\mathcal{NP}$-complete.[2] Thus, it seems justified to settle for the modest goal to find a partial optimum here.

Instead of employing an alternating search on x and β as outlined above, we can locally optimize over (x, β, τ). Note that in this case, the feasible region is generically nonconvex, and there is no guarantee that a local search procedure determines a local optimum or even a feasible solution to problem (5.7). In the next section, we will compare both solution approaches on a set of problem instances.

We close with an example that illustrates model (5.7) and Algorithm 2.

Example 5.3.3. Consider again the deterministic resource allocation problem described in Example 5.2.1. We now assume that the work content of each task $i \in V$ is a uniformly distributed random variable $\widetilde{\omega}_i$ with support $[(1-\zeta)\omega_i, (1+\zeta)\omega_i]$, where ω_i denotes the nominal work content (taken from Example 5.2.1) and $\zeta = 0.2$. For ease of exposition, we assume that the work contents of different project activities are independent.

A uniform distribution with support $[(1-\zeta)\omega_i, (1+\zeta)\omega_i]$ has an expected value of ω_i and a variance of $(\zeta\omega_i)^2/3$. We therefore have $\mu = (2, 5, 1, 4, 3, 1)^\top$, while Σ is given by $\Sigma_{11} \approx 0.053$, $\Sigma_{22} \approx 0.333$, $\Sigma_{33} \approx 0.013$, $\Sigma_{44} \approx 0.213$, $\Sigma_{55} \approx 0.120$, $\Sigma_{66} \approx 0.013$ and $\Sigma_{ij} = 0$ for all $i \neq j$. The project in Example 5.2.1 has activity paths $\overline{\mathcal{P}} = \{P^1, P^2, P^3\}$ with $P^1 = \{1, 2, 4, 6\}$, $P^2 = \{1, 2, 5, 6\}$ and $P^3 = \{1, 3, 5, 6\}$.

[2] Indeed, a procedure that decides local optimality in bilinear problems can be used to verify local optimality in indefinite quadratic problems. The latter problem, however, is known to be $\mathcal{NP}$-complete [HPT00].

For $\alpha = 0.95$, the model (5.7) reads as follows:

$$
\begin{aligned}
\underset{r,x,\beta,\tau}{\text{minimize}} \quad & \tau \\
\text{subject to} \quad & r \in \mathbb{R}^6_+, \quad x \in \mathbb{R}^{12}_+, \quad \beta \in \mathbb{R}^3_+, \quad \tau \in \mathbb{R}_+ \\
& \tau \geq 2r_1 + 5r_2 + 4r_4 + r_6 + \Phi^{-1}(\beta_1) \\
& \quad \times \sqrt{0.053r_1^2 + 0.333r_2^2 + 0.213r_4^2 + 0.013r_6^2},
\end{aligned}
$$

$$
\begin{aligned}
& \tau \geq 2r_1 + 5r_2 + 3r_5 + r_6 + \Phi^{-1}(\beta_2) \\
& \quad \times \sqrt{0.053r_1^2 + 0.333r_2^2 + 0.120r_5^2 + 0.013r_6^2}, \\
& \tau \geq 2r_1 + r_3 + 3r_5 + r_6 + \Phi^{-1}(\beta_3) \\
& \quad \times \sqrt{0.053r_1^2 + 0.013r_3^2 + 0.120r_5^2 + 0.013r_6^2}, \\
& r_1 \left(x_1^1\right)^2 \left(x_1^2\right)^{3/2} \geq 1, \quad r_2 \left(x_2^1\right)^2 \left(x_2^2\right)^{3/2} \geq 1, \quad r_3 \left(x_3^1\right)^2 \left(x_3^2\right)^{3/2} \geq 1, \\
& r_5 \left(x_5^1\right)^2 \left(x_5^2\right)^{3/2} \geq 1, \quad r_6 \left(x_6^1\right)^2 \left(x_6^2\right)^{3/2} \geq 1, \; 2r_4 x_4^1 \geq 1, \quad r_4 x_4^2 \geq 1, \\
& x^1, x^2 \in [(1/4)\mathrm{e}, 2\mathrm{e}], \quad \sum_{i=1}^{6} x_i^1 \leq 6, \quad \sum_{i=1}^{6} x_i^2 \leq 6, \\
& \beta_1 + \beta_2 + \beta_3 \geq 2.95, \quad \beta \in [0, \mathrm{e}].
\end{aligned}
$$

Table 5.1 documents the steps of Algorithm 2 when being applied to this instance. Variables printed in bold are updated in the respective step of the procedure. The algorithm terminates because the improvement of the objective value τ does not exceed a tolerance of 10^{-4}. Note that apart from the increased objective value, the determined solution is very similar to the deterministic resource allocation found in

Table 5.1 Application of Algorithm 2 to the project in Example 5.3.3

x_1^1	x_1^2	x_2^1	x_2^2	x_3^1	x_3^2	x_4^1	x_4^2	x_5^1	x_5^2	x_6^1	x_6^2	β_1	β_2	β_3	τ
0.25	**0.25**	**0.25**	**0.25**	**0.25**	**0.25**	**0.25**	**0.25**	**0.25**	**0.25**	**0.25**	**0.25**	**n/a**	**n/a**	**n/a**	**∞**
0.25	0.25	0.25	0.25	0.25	0.25	0.25	0.25	0.25	0.25	0.25	0.25	**≈ 1**	**0.950**	**≈ 1**	1,483.41
1.20	**1.08**	**1.40**	**1.26**	**0.79**	**0.71**	**0.55**	**1.10**	**1.03**	**0.93**	**1.03**	**0.92**	≈ 1	0.950	≈ 1	8.72
1.20	1.08	1.40	1.26	0.79	0.71	0.55	1.10	1.03	0.93	1.03	0.92	**0.997**	**≈ 1**	**0.953**	8.38
1.21	**1.09**	**1.40**	**1.26**	**0.79**	**0.71**	**0.54**	**1.09**	**1.02**	**0.92**	**1.04**	**0.93**	0.997	≈ 1	0.953	8.37
1.21	1.09	1.40	1.26	0.79	0.71	0.54	1.09	1.02	0.92	1.04	0.93	**0.997**	**≈ 1**	**0.953**	8.37
1.21	**1.09**	**1.40**	**1.26**	**0.79**	**0.71**	**0.54**	**1.09**	**1.02**	**0.92**	**1.04**	**0.93**	0.997	≈ 1	0.953	8.37

The first data row documents the initial solution determined in Step 1. The following rows present the intermediate solutions generated in three consecutive iterations of Steps 2 and 3

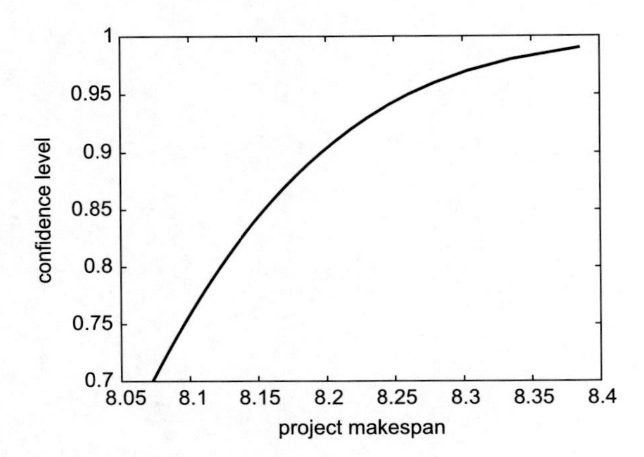

Fig. 5.5 Impact of the confidence level α on the optimal solution to problem (5.7). For higher confidence levels, the estimated makespan increases disproportionately

Example 5.2.1. This is due to the fact that the variance of activity i's duration is chosen to be proportional to the expected value of i's duration. In general, this is not the case, and the determined resource allocations differ significantly. Figure 5.5 shows the impact of the confidence level α on the optimal solution to problem (5.7).

5.4 Numerical Results

In the following, we provide numerical results for the stochastic resource allocation problem (5.4). We do not consider the deterministic model (5.3) for two reasons. Firstly, problem (5.3) is convex and of moderate size and as such, it is clear that it can be solved efficiently even for large projects. Secondly, it is difficult to compare problem (5.3) with other deterministic models (such as the ones discussed in the introduction) which rely on different assumptions.

This section is structured as follows. We start with a comparison of sequential convex and local optimization for solving problem (5.7). We remind the reader that model (5.7) constitutes the approximation of the original resource allocation problem under uncertainty (5.4) obtained by separating the joint chance constraint and approximating the path durations via normal distributions. Afterwards, we compare model (5.7) with alternative approaches to solve problem (5.4). All numerical results are based on averages over 100 randomly generated projects with n activities and $2n$ precedences. The project graphs are constructed with a variant of the deletion method presented in [DDH93]. The work contents follow independent normal or beta distributions with randomly selected parameters. All instances involve two resources, and resource consumption is limited to a third of the sum of upper investment bounds. The activity types (i.e., substitutional or

Table 5.2 Comparison of sequential convex and local optimization

		Instance size n			
		5	10	15	20
Local search procedure	# trials	1.00	1.03	1.04	1.05
	Suboptimality	0.00%	0.00%	0.00%	0.00%
	# iterations	3.00	3.09	3.25	3.98
Sequential convex optimization	Suboptimality (1x)	0.74%	2.96%	3.79%	3.61%
	Suboptimality (10x)	0.36%	1.55%	2.75%	2.79%
	Suboptimality (100x)	0.01%	0.86%	1.94%	2.07%

The table provides the number of optimization runs required to determine a feasible solution (local search procedure) and the number of iterations required to determine a partial optimum (sequential convex optimization). We also record the relative suboptimality of the obtained solutions

limitational) and the parameters $(\underline{c}_i, \overline{c}_i)$, ψ_i, δ_i and γ_i are also chosen randomly. Throughout this section, our goal is to find a resource allocation that minimizes the 0.95-VaR of the uncertain project makespan. All results in this section were obtained with the freely available optimization package Ipopt.[3]

In Sect. 5.3 we discussed two alternative methods for solving problem (5.7): sequential convex optimization (Algorithm 2) determines a partial optimum by solving a series of convex optimization problems, whereas a local search procedure jointly optimizes over all problem variables. Since problem (5.7) is generically nonconvex (see Example 5.3.2), neither approach is guaranteed to provide globally optimal solutions. More so, the local search procedure cannot even guarantee to provide a feasible solution. Table 5.2 compares both approaches on a set of test instances with normally distributed work contents. The quality of the resulting approximate solutions is measured relative to the true global optima, which we determine exactly for these small problem instances ($n \leq 20$) by means of a branch-and-bound algorithm [HPT00]. Table 5.2 reveals that the local search procedure found global optima in all test cases. Although this procedure is more likely to fail on larger problems, it turns out to be very reliable in all considered instances. For sequential convex optimization, we provide the results for a single trial ("1x") and several multi-start ("10x" and "100x") versions. As expected, repeating the search with different start points leads to better solutions. Although sequential convex optimization manages to find good solutions, it is clearly outperformed by the local search procedure. Thus, we will employ the local search procedure in all subsequent tests.

In the remainder of this section, we compare the model (5.7) with three alternative approaches to approximate the original problem (5.4): a nominal problem formulation, a convex approximation via CVaR constraints, and a formulation based on robust optimization. In the nominal problem formulation, the uncertain work contents are replaced with their expected values. The resulting model is a deterministic resource allocation problem of type (5.3). This approach is very

[3]Ipopt homepage: `https://projects.coin-or.org/Ipopt`.

attractive from a computational viewpoint, but it completely ignores the risk inherent to the chosen resource allocation. Nevertheless, nominal formulations are very popular in both theory and practice, and they allow us to quantify the benefits of an honest treatment of uncertainty. As for the CVaR approximation, we replace the joint chance constraint (5.4b) by a related CVaR constraint, which results in a conservative approximation (Sect. 2.2.1). Although the CVaR constraint does not require enumeration of the activity paths, it has no closed-form representation, and we need to employ scenario approximation techniques. In our tests, we approximate the CVaR via 1,000, 2,500 and 5,000 scenarios and a Benders decomposition scheme [Pré95]. As for the approximation based on robust optimization, finally, we use the approach presented in [CSS07]. Since the activity durations fail to be conic functions of the resource investments (in the sense of [CSS07]), we need to enumerate the activity paths in a similar manner as in model (5.7). Contrary to the other formulations, the robust optimization approach is only applicable in the presence of beta-distributed work contents. This is due to the fact that the robust optimization approach in [CSS07] requires all random variables to possess bounded supports.

Our comparison proceeds in two steps. First, we consider instances with normally distributed work contents. It follows from Sect. 5.3 that in this case model (5.7) provides a conservative approximation of the VaR. Afterwards, we compare the formulations on instances with beta-distributed work contents. In this case, model (5.7) does not provide a conservative approximation anymore.

Figure 5.6 and Table 5.3 summarize the results over test instances with normally distributed work contents. The "prediction error" denotes the relative difference between the a priori 0.95-VaR implied by the solutions of the respective optimization models and the a posteriori 0.95-VaR determined by Monte Carlo sampling. We also compare the solutions obtained from the various models in terms of their 0.95-VaR and average makespan, again using Monte Carlo sampling. Both the 0.95-VaR and the average makespan are measured relative to the solution to model (5.7).

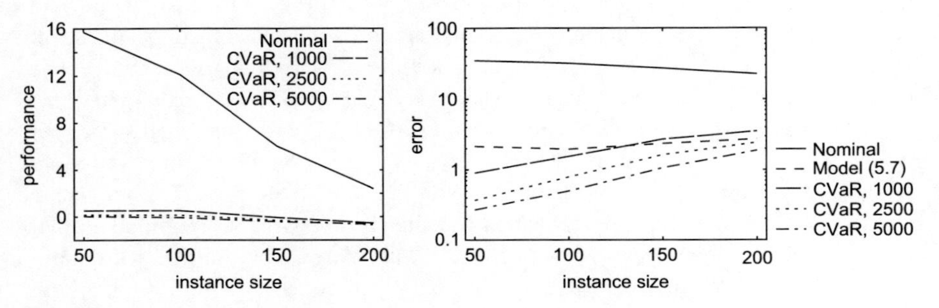

Fig. 5.6 Comparison of model (5.7) with alternative problem formulations for normally distributed work contents. The *left graph* relates the 0.95-VaR of the solutions determined by the alternative formulations to the one obtained from solving model (5.7). The *right graph* shows the relative difference between the estimated and exact 0.95-VaR

Table 5.3 Comparison of model (5.7) with alternative problem formulations for normally distributed work contents

		Instance size n			
		50	100	150	200
Model (5.7)	Prediction error	2.15%	1.96%	2.36%	2.73%
Nominal	Prediction error	35.02%	32.05%	27.34%	22.68%
	0.95-VaR	+15.70%	+12.11%	+6.04%	+2.40%
	Average makespan	+8.45%	+6.03%	+1.07%	−1.44%
CVaR (1000)	Prediction error	0.90%	1.55%	2.71%	3.53%
	0.95-VaR	+0.52%	+0.52%	−0.06%	−0.54%
	Average makespan	+0.55%	+0.42%	−0.36%	−0.37%
CVaR (2500)	Prediction error	0.37%	0.79%	1.63%	2.44%
	0.95-VaR	+0.16%	+0.13%	−0.30%	−0.69%
	Average makespan	+0.05%	+0.16%	−0.51%	−0.60%
CVaR (5000)	Prediction error	0.27%	0.50%	1.06%	1.88%
	0.95-VaR	+0.04%	−0.07%	−0.42%	−0.63%
	Average makespan	−0.14%	−0.09%	−0.71%	−0.58%

"Prediction error" refers to the relative difference between the estimated and exact 0.95-VaR. The 0.95-VaR and average makespan are measured relative to the optimal solution to model (5.7)

The results reveal that the nominal problem grossly underestimates the makespan. This is caused by two factors. Firstly, the nominal problem considers the expected makespan, which is often significantly smaller than the 0.95-VaR. Secondly, by interchanging the maximum and expectation operators in the problem formulation, the nominal model underestimates the expected makespan due to Jensen's inequality, see Sect. 1.1. This underestimation leads to substantial prediction errors and a poor performance of the resulting resource allocations. Indeed, nominal solutions are only acceptable for very large projects, where the assumption of independent work contents makes it increasingly unlikely that the project duration differs significantly from the expected makespan. Note that model (5.7) and the CVaR approximations perform more or less equally well on the considered test instances.

Table 5.4 compares the computational requirements of the considered approaches over test instances with normally distributed work contents. Problem (5.7) needs to be solved only once, but it can involve a large number of activity paths. The table shows, however, that the number of paths remains moderate even for large instances. The CVaR approximations, on the other hand, require the repeated solution of Benders subproblems. It turns out that the number of subproblems increases rapidly with the project size. Thus, the CVaR approximations require substantially more computing resources than the (local) solution of formulation (5.7). This becomes particularly important if some of the considered project resources in model (5.4) are discrete. Note that the nominal model is a deterministic resource allocation problem of type (5.3) and can hence be solved efficiently for all considered sizes.

Table 5.4 Computational requirements of the various problem formulations for normally distributed work contents

	Instance size n			
	50	100	150	200
Model (5.7)	46.34	115.77	189.35	270.96
CVaR (1000)	139.53	283.54	521.78	692.63
CVaR (2500)	141.41	294.06	537.50	678.95
CVaR (5000)	142.56	284.00	522.22	708.35

For model (5.7), the table documents the cardinality of $\overline{\mathcal{P}}$. For the CVaR approximations, the table provides the number of introduced Benders cuts.

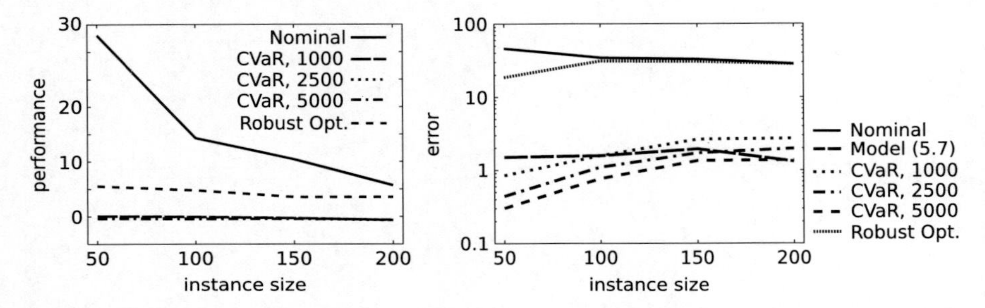

Fig. 5.7 Comparison of model (5.7) with alternative problem formulations for beta-distributed work contents. See Fig. 5.6 for further explanations

Figure 5.7 and Table 5.5 summarize the results for beta-distributed work contents. In this setting, the nominal problem performs even worse than before: both the prediction errors and the 0.95-VaR have deteriorated. Again, model (5.7) and the CVaR approximations perform more of less equally well. It becomes apparent that robust optimization leads to large prediction errors. In contrast to the nominal problem, however, robust optimization overestimates the 0.95-VaR. Although the obtained resource allocations are better than the nominal solutions, they are still substantially worse than the allocations obtained from model (5.7) and the CVaR approximations. Table 5.6 compares the computational requirements of model (5.7) and the CVaR approximations. The results are similar to those of Table 5.4, although the CVaR approximations require slightly more cuts than before. Note that the computational requirements for solving model (5.7) and the robust optimization problem are roughly similar since both formulations scale with the number of activity paths.

In conclusion, although model (5.7) is nonconvex, it seems very well-behaved: in our numerical tests, the model could be solved efficiently and reliably by standard local optimization techniques. Furthermore, the solution quality is comparable to that obtained by convex CVaR approximations of the original model (5.4), even though model (5.7) requires significantly fewer computational resources. The nominal problem and the approximation based on robust optimization can both be

Table 5.5 Comparison of model (5.7) with alternative problem formulations for beta-distributed work contents

		Instance size n			
		50	100	150	200
Model (5.7)	Prediction error	1.51%	1.57%	1.95%	1.31%
Nominal	Prediction error	46.11%	34.45%	32.92%	28.43%
	0.95-VaR	+33.02%	+14.29%	+10.42%	+5.68%
	Average makespan	+15.28%	+5.70%	+2.32%	−0.42%
CVaR (1000)	Prediction error	0.85%	1.57%	2.64%	2.71%
	0.95-VaR	+0.08%	−0.06%	−0.28%	−0.66%
	Average makespan	+1.59%	+0.74%	+0.23%	−0.39%
CVaR (2500)	Prediction error	0.44%	1.10%	1.71%	1.97%
	0.95-VaR	−0.25%	−0.37%	−0.44%	−0.67%
	Average makespan	+1.19%	+0.42%	+0.17%	−0.10%
CVaR (5000)	Prediction error	0.30%	0.77%	1.35%	1.37%
	0.95-VaR	−0.37%	−0.44%	−0.40%	−0.60%
	Average makespan	+1.02%	+0.31%	+0.11%	+0.07%
Robust optimization	Prediction error	18.70%	31.10%	30.61%	28.26%
	0.95-VaR	+5.53%	+4.78%	+3.56%	+3.57%
	average makespan	+7.91%	+4.28%	+3.14%	+2.35%

See Table 5.3 for further explanations

Table 5.6 Computational requirements of the various problem formulations for beta-distributed work contents

	Instance size n			
	50	100	150	200
Model (5.7)	46.26	114.87	190.58	266.43
CVaR (1000)	139.43	319.78	574.38	771.26
CVaR (2500)	139.14	318.89	587.91	808.19
CVaR (5000)	140.41	318.62	582.09	794.38

See Table 5.4 for further explanations

solved very efficiently, but they lead to poor makespan estimates and thus suggest severely suboptimal resource allocations.

5.5 Extensions

In this section, we first illustrate how one can robustify model (5.7) against uncertainty in the first and second moments of the work contents. Afterwards, we present an iterative solution procedure for model (5.7) which applies to projects with large numbers of activity paths. In the following, we abbreviate the β_l-quantile of the duration of activity path $P^l \in \overline{\mathcal{P}}$ by

$$q_l(x, \beta_l; \mu, \Sigma) = \mu^\top \varrho_l(x) + \Phi^{-1}(\beta_l)\sqrt{\varrho_l(x)^\top \Sigma\, \varrho_l(x)}.$$

5.5.1 Moment Ambiguity

The stochastic resource allocation model (5.7) minimizes the α-VaR of the project makespan and therefore hedges against the uncertainty underlying the factual work contents. However, the model assumes rather detailed knowledge about the nature of this uncertainty since it requires a precise specification of its first two moments. Here, we relax this assumption and require instead that these moments are merely known to be contained in the set $\mathcal{U} = \mathcal{U}_\mu \times \mathcal{U}_\Sigma$ with

$$\mathcal{U}_\mu = \left\{\mu \in \mathbb{R}^n_+ \,:\, \mu = \mu_0 + w_\mu \bullet \widehat{\mu},\, \left\|\widehat{\mu}\right\|_2 \leq 1,\, \widehat{\mu} \in \mathbb{R}^n\right\}$$

$$\text{and} \quad \mathcal{U}_\Sigma = \left\{\Sigma \in \mathbb{S}^n_+ \,:\, \Sigma = \Sigma_0 + W_\Sigma \bullet \widehat{\Sigma},\, \left\|\widehat{\Sigma}\right\|_2 \leq 1,\, \widehat{\Sigma} \in \mathbb{S}^n_+\right\},$$

where $\mu_0 \in \mathbb{R}^n_+$ and $\Sigma_0 \in \mathbb{S}^n_+$. The operator "$\bullet$" denotes the element-wise (Hadamard) product, while $\mathbb{S}^n_+$ denotes the subspace of symmetric and positive semidefinite matrices in $\mathbb{R}^{n \times n}$. The parameters μ_0 and Σ_0 can be interpreted as nominal values, while $w_\mu \in \mathbb{R}^n_+$ and $W_\Sigma \in \mathbb{R}^{n \times n}_+$ represent "degrees of ambiguity".

In the spirit of robust optimization (Sect. 2.2.2), our goal is to minimize the worst-case α-VaR of the project makespan under the assumption that the true moments (μ, Σ) can be any element of $\mathcal{U}$. This can be expressed as

$$\min_{\substack{x \in X,\\ \beta \in B}} \max_{P^l \in \mathcal{P}} \underbrace{\max_{(\mu,\Sigma) \in \mathcal{U}} q_l(x, \beta_l; \mu, \Sigma)}_{\varphi_l(x,\beta_l)}.$$

Due to the separability of $\mathcal{U}$ with respect to the first and second moments, the worst-case α-VaR is representable as

$$\begin{aligned}
\varphi_l(x, \beta_l) &= \max_{(\mu,\Sigma) \in \mathcal{U}} \left\{\mu^\top \varrho_l(x) + \Phi^{-1}(\beta_l) \sqrt{\varrho_l(x)^\top \Sigma \, \varrho_l(x)}\right\} \\
&= \max_{\mu \in \mathcal{U}_\mu} \left\{\mu^\top \varrho_l(x)\right\} + \Phi^{-1}(\beta_l) \max_{\Sigma \in \mathcal{U}_\Sigma} \left\{\sqrt{\varrho_l(x)^\top \Sigma \, \varrho_l(x)}\right\}.
\end{aligned}$$

The first maximization term reduces to

$$\begin{aligned}
\max_{\mu \in \mathcal{U}_\mu} \left\{\mu^\top \varrho_l(x)\right\} &= (\mu_0)^\top \varrho_l(x) + \max_{\substack{\left\|\widehat{\mu}\right\|_2 \leq 1,\\ \mu_0 + \widehat{\mu} \geq 0}} \left(w_\mu \bullet \widehat{\mu}\right)^\top \varrho_l(x) \\
&= (\mu_0)^\top \varrho_l(x) + \max_{\substack{\left\|\widehat{\mu}\right\|_2 \leq 1,\\ \mu_0 + \widehat{\mu} \geq 0}} \widehat{\mu}^\top \left[w_\mu \bullet \varrho_l(x)\right] \\
&= (\mu_0)^\top \varrho_l(x) + \left\|w_\mu \bullet \varrho_l(x)\right\|_2 .
\end{aligned}$$

Concerning the last identity, note that all components of $w_\mu \bullet \varrho_l(x)$ are nonnegative, and hence $\widehat{\mu} \geq 0$ and $\mu_0 + \widehat{\mu} \geq 0$ are vacuously satisfied at optimality. By applying similar transformations as described in Proposition 5.3.2, the last term can be expressed by convex constraints.

Likewise, one can show that the Σ-term reduces to

$$\max_{\Sigma \in \mathcal{U}_\Sigma} \left\{ \sqrt{\varrho_l(x)^\top \, \Sigma \, \varrho_l(x)} \right\} = \sqrt{\varrho_l(x)^\top \, \Sigma_0 \varrho_l(x) + \left\| W_\Sigma \bullet \left[\varrho_l(x) \varrho_l(x)^\top \right] \right\|_2}.$$

For $\Sigma_0 \succeq 0$ (see Sect. 5.3), the latter term can be expressed by convex constraints, too. Hence, the convexity properties of model (5.7) are preserved when the moments of the work contents are ambiguous, and the increase in model size is moderate. This contrasts with the optimization of CVaR under distributional ambiguity, which is considerably more involved [PW07, ZF09].

5.5.2 *Iterative Path Selection Procedure*

The runtime behavior of the stochastic resource allocation model (5.7) depends on the number of activity paths in $\overline{\mathcal{P}}$. Recall that $\overline{\mathcal{P}}$ has been of moderate size in all of our numerical tests (see Sect. 5.4). However, the number of activity paths can be exponential in the size of the project graph, see Chap. 6. In this section we present an iterative solution procedure for problem (5.7) based on the principles of semi-infinite programming. The outline of the procedure is described in Algorithm 3.

Algorithm 3 Iterative path selection procedure for model (5.7).

---

1. Initialize $\mathcal{L}$ as a (nonempty) subset of $\overline{\mathcal{P}}$. Choose $\epsilon \in (0, (1-\alpha)/\left|\overline{\mathcal{P}} \setminus \mathcal{L}\right|)$.
2. Determine a feasible (possibly suboptimal) solution (x^*, β^*) to

$$\begin{aligned} \underset{x,\beta}{\text{minimize}} \quad & \max_{P^l \in \mathcal{L}} \; q_l(x, \beta_l; \mu, \Sigma) \\ \text{subject to} \quad & x \in \mathbb{R}_+^{mn}, \;\; \beta \in \mathbb{R}_+^{|\mathcal{L}|} \\ & \sum_{P^l \in \mathcal{L}} \beta_l \geq \alpha + (\left|\overline{\mathcal{P}}\right| - 1) - \left|\overline{\mathcal{P}} \setminus \mathcal{L}\right| (1 - \epsilon), \\ & x \in X, \;\; \beta \in [0, \mathrm{e}]\,. \end{aligned}$$

   Let τ^* denote the resulting objective value.
3. Check whether there is a path $P^s \in \overline{\mathcal{P}} \setminus \mathcal{L}$ with $q_s(x^*, 1-\epsilon; \mu, \Sigma) > \tau^*$. If this is the case, then add one such path to $\mathcal{L}$ and return to Step 2. Otherwise, stop: x^* represents the best resource allocation found.

Step 1 initializes $\mathcal{L}$, the subset of activity paths $P^l \in \overline{\mathcal{P}}$ which are currently considered. It also assigns a value to ϵ, the probability that is assigned to the paths in $\overline{\mathcal{P}} \setminus \mathcal{L}$ which are not (yet) considered. In *Step 2*, model (5.7) is solved for the activity paths $P^l \in \mathcal{L}$. Note that the first constraint implicitly assigns a probability of $1-\epsilon$ to every path in $\overline{\mathcal{P}} \setminus \mathcal{L}$. *Step 3* checks whether there is a path in $\overline{\mathcal{P}} \setminus \mathcal{L}$ whose $(1-\epsilon)$-duration quantile exceeds the α-VaR determined in the previous step. If this is the case, then one such path is added to $\mathcal{L}$, and the procedure iterates. Otherwise, the procedure terminates. We will present a strategy to determine a suitable path in $\overline{\mathcal{P}} \setminus \mathcal{L}$ below.

If the subproblem in Step 2 is infeasible, then $X = \emptyset$ and problem (5.7) does not possess a feasible solution. For any $\epsilon \in (0, (1-\alpha)/\left|\overline{\mathcal{P}} \setminus \mathcal{L}\right|)$, the final resource allocation obtained by Algorithm 3 is feasible in problem (5.7), and τ^* represents a conservative estimate of its objective value. Note that for a fixed ϵ, only a near-optimal solution is determined if $\mathcal{L} \neq \overline{\mathcal{P}}$ at termination. This statement is true even if the subproblems arising in Step 2 are solved to global optimality. Indeed, a better "probability arrangement" can potentially be obtained by assigning $\beta_l > 1-\epsilon$ to paths $P^l \in \overline{\mathcal{P}} \setminus \mathcal{L}$. This is not restrictive for practical applications, however, since optimization algorithms typically require an upper bound strictly below 1 for β_l, $P^l \in \overline{\mathcal{P}}$, anyway (since $\Phi(\beta_l) \longrightarrow \infty$ for $\beta_l \longrightarrow 1$). For any given value of ϵ, let $x(\epsilon)$ and $f(\epsilon)$ denote any final resource allocation and its objective value, respectively, that are determined by Algorithm 3 when solving the subproblems to global optimality. One can show that the sequence $\{f(\epsilon)\}_{\epsilon \longrightarrow 0}$ converges monotonically to the optimal objective value of problem (5.7). Furthermore, every accumulation point of $\{x(\epsilon)\}_{\epsilon \longrightarrow 0}$ constitutes a globally optimal resource allocation for problem (5.7).

Note that in the third step, we have to examine a potentially large number of paths $P^s \in \overline{\mathcal{P}} \setminus \mathcal{L}$. We can obtain an upper bound on the $(1-\epsilon)$-duration quantile of path $P^s \in \overline{\mathcal{P}} \setminus \mathcal{L}$ as follows:

$$
\begin{aligned}
q_s(x^*, 1-\epsilon; \mu, \Sigma) &= \mu^\top \varrho_s(x^*) + \Phi^{-1}(1-\epsilon)\sqrt{\varrho_s(x^*)^\top \Sigma\, \varrho_s(x^*)} \\
&\leq \mu^\top \varrho_s(x^*) + \Phi^{-1}(1-\epsilon)\sqrt{\sum_{i \in P^s} \eta_i} \quad \text{with} \\
\eta_i &= \left[\max_{\substack{P^l \in \overline{\mathcal{P}}: \\ i \in P^l}} \sum_{j \in P^l} \sigma_{ij} / \left[\rho_i(x_i^*)\rho_j(x_j^*)\right] \right]^+ \\
&\leq \sum_{i \in P^s} \phi_i \quad \text{with} \quad \phi_i = \mu_i / \rho_i(x_i^*) + \Phi^{-1}(1-\epsilon)\sqrt{\eta_i}.
\end{aligned}
$$

Here, we use the abbreviation $[x]^+ = \max\{x, 0\}$. The first inequality holds because $i \in P^s$ implies that $P^s \in \left\{P^l \in \overline{\mathcal{P}} : i \in P^l\right\}$. The second inequality follows from the fact that the 2-norm of a vector is no larger than its 1-norm and $\eta_i \geq 0$. Note that η_i (and hence, ϕ_i) can be determined in polynomial time (relative to the size of the project graph) for all $i \in V$. The described upper bound allows us to construct a deterministic project with durations ϕ_i for $i \in V$. Every path duration in this project yields an upper bound on the $(1-\epsilon)$-duration quantile of the respective path in model (5.7). Thus, we can use techniques for determining the κ largest paths in a directed, acyclic graph to obtain candidates paths $P^s \in \overline{\mathcal{P}} \setminus \mathcal{L}$ for inclusion in $\mathcal{L}$. In particular, we can stop the search in Step 3 once we have examined all paths $P^s \in \overline{\mathcal{P}} \setminus \mathcal{L}$ with $\sum_{i \in P^s} \phi_i > \tau^*$. A method for determining the κ largest paths of a project graph $G = (V, E)$ in time $\mathcal{O}(|E| + \kappa\,|V|)$ is presented in [Epp94].

We will refine Algorithm 3 in the next chapter, where we use a variant of this algorithm to generate convergent lower bounds on the optimal objective value of two-stage robust resource allocation problems. In that chapter, we will also provide a numerical example of the algorithm.

5.6 Conclusion

Resource allocation problems constitute a vital class of project scheduling problems. The first part of this chapter presented a deterministic multi-resource allocation model that is convex and that hence scales to large problem sizes. Resource allocation models have to stipulate functional relations between resource investments and task durations. The considered model employs production functions from microeconomic theory, which lead to intuitively appealing duration functions that are amenable to economic interpretation.

The second part of this chapter discussed an extension of the deterministic resource allocation model that accommodates uncertainty. The formulation assumes knowledge about the first two moments of the uncertain parameters and optimizes the α-VaR of the project makespan. Although VaR is a nonconvex risk measure, we showed that the specific properties of project scheduling problems enable us to approximately optimize it very efficiently. Furthermore, the resulting model readily accommodates distributional ambiguity. This is useful in project scheduling, because the moments of the uncertain parameters are often unknown due to the lack of historical data.

While the presented resource allocation model seems to be primarily suitable for project scheduling problems, the VaR approximation readily applies to other application areas of temporal networks such as process scheduling [JM99] and digital circuit design [KBY$^+$07]. It would therefore be instructive to apply variants of the approximate problem formulation (5.7) to models in these application areas as well.

Chapter 6
Minimization of the Worst-Case Makespan

6.1 Introduction

In this chapter we study a robust resource allocation problem that minimizes the worst-case makespan. As in the previous chapters, we assume that the resource allocation is a here-and-now decision, whereas the task start times are modeled as a wait-and-see decision that may depend on random parameters affecting the task durations. In the terminology of Sect. 2.2.2, we therefore study a two-stage robust optimization problem. In contrast to its stochastic counterpart, the complexity of the robust resource allocation problem has been unknown for a long time [Hag88]. The majority of the solution approaches presented in the literature determine suboptimal solutions without bounding the incurred optimality gap. In this chapter, we show that the robust resource allocation problem is $\mathcal{NP}$-hard, which explains the lack of exact solution approaches in the literature. We then present two hierarchies of approximate problems that provide convergent lower and upper bounds on the optimal value of the original problem. The upper bounds correspond to feasible allocations whose objective values are bracketed by the bounds. Hence, we obtain a sequence of feasible allocations that are asymptotically optimal and whose optimality gaps can be quantified at any time.

There are three robust resource allocation problems in temporal networks that directly relate to the problem considered in this chapter. A production scheduling problem that minimizes the worst-case makespan under uncertain processing times, product demands and market prices is proposed in [JLF07, LJF04]. The decision maker can influence the makespan by choosing a processing sequence and assigning resources to the individual processing steps, and the optimal process start times are approximated by constant decision rules. A robust variant of the time/cost trade-off problem in project scheduling is discussed in [CSS07]. Assuming that the durations of the project activities are uncertain, this model determines a resource allocation that minimizes the worst-case makespan. To obtain a tractable optimization problem, the optimal task start times are approximated by affine decision rules. A related time/cost trade-off problem is studied in [CGS07], where

W. Wiesemann, *Optimization of Temporal Networks under Uncertainty*, Advances in Computational Management Science 10,
DOI 10.1007/978-3-642-23427-9_6, © Springer-Verlag Berlin Heidelberg 2012

the resource allocation for a specific activity is allowed to adapt to all uncertain parameters that have been observed until the respective task start time. Affine decision rules are used to obtain a tractable approximation for the problem. We will review decision rules in Sect. 6.2.2.

Research in the wider area of robust network optimization started with the seminal paper [BS03], which develops solution techniques for single-stage robust network flow problems. In recent years, several two-stage robust network optimization problems have been solved under the name of recoverable robust optimization. In [LLMS09], a railway scheduling problem is considered which selects a here-and-now timetable that can be made feasible for a range of train delays in the second stage. A two-stage robust freight transportation problem is studied in [EMS09]. This model determines a here-and-now repositioning plan for empty containers that can be recovered for a range of supply and demand scenarios in the second stage. In both papers, tractable optimization problems are derived through carefully chosen problem reformulations. In [AZ07, OZ07], a two-stage robust network optimization problem is proposed which treats the network design as a here-and-now decision, while the network flows are modeled as wait-and-see decisions that are chosen after the uncertain parameters have been observed. Recently, approximation algorithms have been developed for two-stage robust combinatorial problems [FJMM07, KKMS08]. Here, a feasible solution to the combinatorial problem has to be found for any possible realization of the random parameters. Since the second stage decision incurs a higher cost, there is a trade-off between over-protection in the first stage and a costly recovery in the second stage. Finally, there is an extensive literature on network problems that optimize the worst-case regret, see [Ave01].

The remainder of this chapter is organized as follows. In the next section, we define the robust resource allocation problem. After a review of popular approximations for the problem, we show that the robust resource allocation problem is generically $\mathcal{NP}$-hard. In Sect. 6.3 we discuss a path-wise formulation that provides the basis for the solution technique. This formulation follows the spirit of the stochastic resource allocation problem presented in the previous chapter. In Sects. 6.4 and 6.5 we present families of optimization problems that provide convergent lower and upper bounds, respectively. Section 6.6 discusses the results of a numerical evaluation on randomly generated test instances, and Sect. 6.7 applies the bounding scheme to VLSI design. We conclude in Sect. 6.8.

In addition to the notation introduced in Sect. 1.3, this chapter uses the following convention. For a set $A \subseteq \{1, \ldots, n\}$, we denote by $\mathbb{I}_A$ the n-dimensional vector with $(\mathbb{I}_A)_i = 1$ if $i \in A$ and $(\mathbb{I}_A)_i = 0$ otherwise. As we will see shortly, this allows us to express the sum of task durations on a network path $P \subseteq V$ as the inner product between the indicator vector $\mathbb{I}_P$ of the path and the vector of all task durations.

6.2 Robust Resource Allocations

We first define the robust resource allocation problem that we consider in this chapter. We then review how decision rules can be applied to obtain a tractable approximation for this problem. In Sect. 6.2.3 we analyze the complexity of the robust resource allocation problem.

6.2.1 *The Robust Resource Allocation Problem*

We assume that the structure of the temporal network (i.e., V and E) is deterministic, whereas the task durations are uncertain, see Sect. 2.3. We model the duration of task $i \in V$ by a continuous function $d_i : X \times \Xi \mapsto \mathbb{R}_+$ that maps resource allocations $x \in X$ and realizations of the uncertain parameters $\xi \in \Xi$ to nonnegative durations. Examples of duration functions are discussed in the previous chapter. We assume that both X, the set of admissible resource allocations, and Ξ, the support of the uncertain parameters, are nonempty and compact subsets of finite-dimensional spaces. Having in mind the application areas outlined in Sect. 1.1, we assume that ξ cannot be observed directly, but that it can only be gradually inferred from the durations of completed tasks, see Sect. 2.3. In strategic decision problems, Ξ is sometimes specified as a discrete set of rival scenarios (e.g., different forecasts of market developments). We will see that under rather general convexity assumptions, robust allocation problems that minimize the worst-case makespan over finite discrete supports Ξ can be formulated as explicit convex programs. Often, however, Ξ is better described by a set of infinite cardinality, such as an ellipsoid around a nominal parameter vector. In this chapter, we focus on uncertainty sets that are of infinite cardinality but specific structure.

We define the *robust resource allocation problem on temporal networks* as

$$\min_{x \in X} \max_{\xi \in \Xi} \min_{y \in Y(x,\xi)} \{y_n + d_n(x;\xi)\}, \qquad (\mathcal{RTN})$$

where

$$Y(x,\xi) = \left\{y \in \mathbb{R}^n_+ : y_j \geq y_i + d_i(x;\xi) \;\; \forall (i,j) \in E\right\}. \qquad (6.1)$$

For $x \in X$ and $\xi \in \Xi$, $Y(x,\xi)$ denotes the set of admissible start time vectors for the network tasks. $\mathcal{RTN}$ is a two-stage robust optimization problem: the uncertain parameters $\xi \in \Xi$ are revealed after the allocation x has been chosen, but before the task start times y have been decided upon. Hence, we are interested in a static resource allocation which cannot be adapted once information about ξ becomes available. We have already mentioned the reasons for our interest in static allocations in the previous chapter: resource allocations are frequently required to be static due to the inflexibility of resources and limitations of the manufacturing process, or to enhance the planning security and the compatibility with concurrent operations outside the scope of the model. Even in situations where recourse decisions are principally possible, static allocations might be preferable to ensure computational tractability [GG06, JWW98]. To illustrate the importance of static resource allocations, consider the gate sizing problem outlined in Sect. 1.1. The gate sizes have to be chosen before the impact of process deviations is known. Hence, only static allocations are meaningful in digital circuit design. Unlike the resource allocation x, the task start times y may typically depend on the available knowledge about ξ. Note that every component of y is chosen after *all* uncertain parameters are revealed, which seems to violate non-anticipativity [RS03]: the uncertain parameters are revealed gradually when tasks are completed, and y_j, $j \in V$, must only depend on information that is available at the time when task j is started. The justification for the chosen two-stage structure is the same as in the previous chapters. The early start schedule $y^* : X \times \Xi \mapsto \mathbb{R}^n_+$ with $y_1^*(x, \xi) = 0$ and

$$y_j^*(x, \xi) = \max_{i \in V} \{y_i^*(x, \xi) + d_i(x; \xi) \; : \; (i, j) \in E\} \quad \text{for all } j \in V \setminus \{1\}$$

is non-anticipative since the task start times only depend on the completion times of predecessor tasks. Moreover, since the makespan is a non-decreasing function of the task start times, the early start schedule is also optimal. Hence, if a solution to $\mathcal{RTN}$ employs an anticipative start time schedule y, then we can replace it with the corresponding (non-anticipative) early start schedule without sacrificing optimality.

The robust resource allocation problem treated in this chapter has relevance in all application areas outlined in Sect. 1.1. The solution approach presented in this chapter is also suited for several variants of $\mathcal{RTN}$, such as multi-objective problems that contain the makespan as one of several goals and problems with makespan restrictions as side constraints. We will see an example of such an extension in the case study in Sect. 6.7.

6.2.2 Decision Rule Approximations

$\mathcal{RTN}$ constitutes a min–max–min problem with coupled constraints and is as such not amenable to standard optimization techniques. Most existing solution approaches rely on the following observation to obtain a tractable approximation to $\mathcal{RTN}$.

Observation 6.2.1 *For the robust resource allocation problem $\mathcal{RTN}$, we have*

$$\min_{x \in X} \max_{\xi \in \Xi} \min_{y \in Y(x, \xi)} \{y_n + d_n(x; \xi)\} = \min_{\substack{x \in X, \\ y \in \mathcal{Y}(x)}} \max_{\xi \in \Xi} \{y_n(\xi) + d_n(x; \xi)\}, \tag{6.2a}$$

where for $x \in X$,

$$\mathcal{Y}(x) = \left\{(y : \Xi \mapsto \mathbb{R}^n_+) \; : \; y(\xi) \in Y(x, \xi) \;\; \forall \xi \in \Xi\right\}. \tag{6.2b}$$

For a resource allocation $x \in X$, $\mathcal{Y}(x)$ denotes the space of all functions on Ξ that map parameter realizations to feasible start time vectors for the tasks.

Note that the identity (6.2a) holds regardless of the properties of X and d because $\mathcal{Y}(x)$ does not impose any structure on the functions y (such as measurability). Observation 6.2.1 allows us to reduce the min–max–min problem $\mathcal{RTN}$ to a min–max problem at the cost of augmenting the set of first-stage decisions. We have already encountered this transformation in Sect. 2.2.2 when we discussed generic two-stage robust optimization problems. A function y is called a *decision rule* because it specifies the second-stage decision as a function of the uncertain parameters. Note that the choice of an appropriate decision rule is part of the first-stage decision. Since $\mathcal{Y}(x)$ constitutes a function space, further assumptions are required to ensure solvability. For example, if Ξ contains finitely many scenarios, $\Xi = \{\xi^1, \ldots, \xi^L\}$, then $\mathcal{Y}(x)$ is isomorphic to a subset of $\mathbb{R}_+^{Ln}$ and we can reformulate $\mathcal{RTN}$ as

$$\min_{\substack{x \in X, \\ y \in \mathbb{R}_+^{Ln}}} \left\{ \max_{l=1,\ldots,L} \{y_n^l + d_n(x, \xi^l)\} : y_j^l \geq y_i^l + d_i(x; \xi^l) \;\; \forall l = 1, \ldots, L, \; (i, j) \in E \right\}.$$

This problem is convex if X is convex and d is convex in its first component for all $\xi^l \in \Xi$. Similar finite-dimensional problems arise when a semi-infinite programming algorithm is used to solve $\mathcal{RTN}$ with an uncertainty set of infinite cardinality [HK93]. This approach, however, would only provide lower bounds on the optimal value of $\mathcal{RTN}$, and it is not clear how to efficiently obtain upper bounds.[1] Furthermore, one would not be able to exploit structural properties of Ξ and d beyond convexity. Finally, the number of constraints and variables grows with L, which itself is likely to become large for tight approximations.

Due to the absence of standard optimization techniques for the solution of $\mathcal{RTN}$ when Ξ has infinite cardinality, one commonly settles for feasible but suboptimal solutions. These are obtained from conservative approximations of $\mathcal{RTN}$ that restrict the set of admissible second-stage decisions. For example, it has been suggested in [LJF04] to restrict $\mathcal{Y}$ to *constant decision rules*, that is, to

$$\mathcal{Y}^0(x) = \{y \in \mathcal{Y}(x) \, : \, \exists \gamma \in \mathbb{R}^n \text{ such that } y(\xi) = \gamma \;\; \forall \xi \in \Xi\} \quad \text{for } x \in X.$$

In this case, $\mathcal{RTN}$ is equivalent to

$$\begin{aligned} &\min_{\substack{x \in X, \\ y \in \mathcal{Y}^0(x)}} \max_{\xi \in \Xi} \{y_n(\xi) + d_n(x; \xi)\} \\ &= \min_{\substack{x \in X, \\ \gamma \in \mathbb{R}_+^n}} \left\{ \max_{\xi \in \Xi} \{\gamma_n + d_n(x; \xi)\} : \gamma_j \geq \gamma_i + d_i(x; \xi) \;\; \forall \xi \in \Xi, \; (i, j) \in E \right\} \end{aligned}$$

[1]As we will see in Sect. 6.2.3, evaluating the worst-case makespan of the optimal second-stage policy in $\mathcal{RTN}$ constitutes a difficult problem even for fixed $x \in X$.

$$= \min_{\substack{x \in X, \\ y \in \mathbb{R}^n_+}} \left\{ y_n + \max_{\xi \in \Xi} \{d_n(x;\xi)\} \ : \ y_j - y_i \geq \max_{\xi \in \Xi} \{d_i(x;\xi)\} \ \ \forall (i,j) \in E \right\}.$$

The tractability of this problem is determined by the properties of X and the functions $\max_{\xi \in \Xi} \{d_i(x;\xi)\}$ for $i \in V$. For general Ξ and d the problem can be formulated as a semi-infinite program [HK93]. For specific choices of Ξ and d, robust optimization techniques can be used to obtain equivalent (or approximate) explicit reformulations [BS06, BTGN09]. Although they are computationally attractive, constant decision rules can result in poor approximations of the optimal second-stage policies and – as a consequence – the optimal resource allocations.

Example 6.2.1. Consider the temporal network $G = (V, E)$ with tasks $V = \{1, \ldots, n\}$ and precedence relations $E = \{(i, i+1) : 1 \leq i < n\}$. Let the uncertainty set be defined as $\Xi = \left\{\xi \in \mathbb{R}^n_+ : \mathrm{e}^\top \xi \leq 1\right\}$ and the (decision-independent) task durations as $d_i(x;\xi) = \xi_i$ for $i \in V$. The optimal second-stage policy incurs a worst-case makespan of 1, whereas the restriction to constant decision rules results in a worst-case makespan of n.

In order to improve on the approximation quality of constant decision rules, it has been suggested in [BTGN09, CSS07] to approximate $\mathcal{Y}(x)$ by the set of *affine decision rules*: for $x \in X$ and $\Xi \subseteq \mathbb{R}^k$, we define

$$\mathcal{Y}^1(x) = \left\{ y \in \mathcal{Y}(x) \ : \ \exists \Gamma \in \mathbb{R}^{n \times k}, \gamma \in \mathbb{R}^n \ \text{ such that } \ y(\xi) = \Gamma\xi + \gamma \ \ \forall \xi \in \Xi \right\}.$$

Under this approximation, $\mathcal{RTN}$ reduces to

$$\min_{\substack{x \in X, \\ y \in \mathcal{Y}^1(x)}} \max_{\xi \in \Xi} \{y_n(\xi) + d_n(x;\xi)\}$$

$$= \min_{\substack{x \in X, \\ \Gamma \in \mathbb{R}^{n \times k}, \\ \gamma \in \mathbb{R}^n}} \left\{ \gamma_n + \max_{\xi \in \Xi} \left\{ \Gamma_n^\top \xi + d_n(x;\xi) \right\} \ : \ (\Gamma, \gamma) \in \mathcal{S}_+ \cap \mathcal{S}_E(x) \right\}$$

with

$$\begin{aligned} \mathcal{S}_+ &= \{(\Gamma, \gamma) \ : \ \Gamma\xi + \gamma \geq 0 \ \ \forall \xi \in \Xi\} \\ &= \left\{ (\Gamma, \gamma) \ : \ \gamma_i \geq \max_{\xi \in \Xi} \left\{ -\Gamma_i^\top \xi \right\} \ \ \forall i \in V \right\} \end{aligned}$$

and
$$\begin{aligned} \mathcal{S}_E(x) &= \left\{ (\Gamma, \gamma) : \Gamma_j^\top \xi + \gamma_j \geq \Gamma_i^\top \xi + \gamma_i + d_i(x;\xi) \quad \forall \xi \in \Xi, \ (i,j) \in E \right\} \\ &= \left\{ (\Gamma, \gamma) : \gamma_j - \gamma_i \geq \max_{\xi \in \Xi} \left\{ (\Gamma_i - \Gamma_j)^\top \xi + d_i(x;\xi) \right\} \ \forall (i,j) \in E \right\}. \end{aligned}$$

Here, $\Gamma_i^\top$ denotes the ith row of matrix Γ. As in the case of constant decision rules, this model can be solved via semi-infinite programming, and under certain

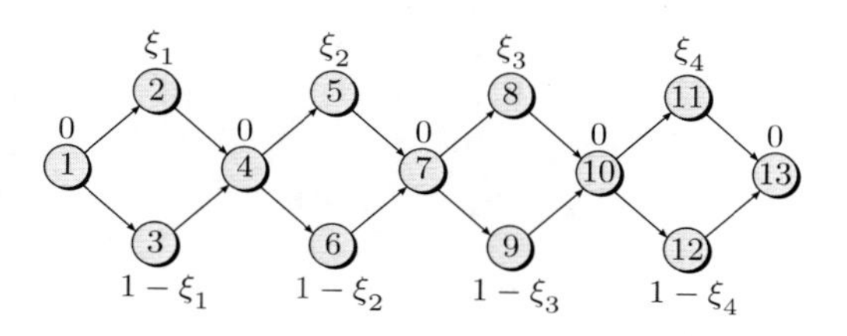

Fig. 6.1 Example temporal network that illustrates the suboptimality of affine decision rules. The graph visualizes the network structure for $k = 4$. The task durations (next to the nodes) are defined in the text

conditions we can employ robust optimization techniques to obtain explicit reformulations. Much like constant decision rules, however, affine decision rules can lead to poor approximations of $\mathcal{RTN}$.

Example 6.2.2. Consider the class of temporal networks illustrated in Fig. 6.1. For $k \in \mathbb{N}$, the network structure is given by $V = \{1, \ldots, 3k + 1\}$ and

$$E = \{(3l + 1, 3l + p), (3l + p, 3l + 4) \ : \ 0 \leq l < k, \ p = 2, 3\}.$$

Let $d_{3l+2}(x; \xi) = \xi_{l+1}$ and $d_{3l+3}(x; \xi) = 1 - \xi_{l+1}$ for $0 \leq l < k$, while the remaining task durations are zero. For $\Xi = \left\{\xi \in \mathbb{R}^k_+ \ : \ \|\xi - (1/2)\,\mathrm{e}\|_1 \leq 1/2\right\}$, the optimal second-stage policy leads to a worst-case makespan of $(k + 1)/2$. For $0 \leq l < k$, we obtain $y_{3l+4}(\xi) \geq y_{3l+1}(\xi) + \max\{\xi_{l+1}, 1 - \xi_{l+1}\}$ for all $\xi \in \Xi$. In particular, this inequality holds for $\xi \in \{(1/2)\,\mathrm{e} \pm (1/2)\,\mathrm{e}_{l+1}\}$, where e_{l+1} denotes the $(l+1)$th vector of the standard basis in $\mathbb{R}^k$. If we restrict y to be affine in ξ, then the previous observation implies that $y_{3l+4}(\xi) \geq y_{3l+1}(\xi) + 1$ for $\xi = (1/2)\,\mathrm{e} \in \Xi$ and

$$y_{3k+1}(\xi) \geq y_{3k-2}(\xi) + 1 \geq \ldots \geq y_1(\xi) + k \geq k \quad \text{for } \xi = (1/2)\,\mathrm{e}.$$

Here, the last inequality holds by nonnegativity of y. Thus, the restriction to affine decision rules results in a worst-case makespan of at least k.

Recently, the use of piecewise affine decision rules has been advocated to overcome some of the deficiencies of affine decision rules [CSSZ08]. Nevertheless, Examples 6.2.1 and 6.2.2 show that the existing solution approaches for $\mathcal{RTN}$ can lead to poor approximations of the optimal decisions. This is supported by our numerical results in Sect. 6.6. In the next section, we show that $\mathcal{RTN}$ constitutes a difficult optimization problem, which explains the lack of exact solution procedures in the literature.

6.2.3 Complexity Analysis

It is clear that $\mathcal{RTN}$ is difficult to solve if we impose no further regularity conditions beyond compactness of X and Ξ. In the following, we show that evaluating the worst-case makespan of the optimal second-stage policy constitutes an $\mathcal{NP}$-complete problem even when the resource allocation $x \in X$ is fixed, while Ξ and d have "simple" descriptions. This implies that $\mathcal{RTN}$ is $\mathcal{NP}$-hard since we can restrict X to a singleton and thus obtain a procedure that evaluates the worst-case makespan of the optimal second-stage policy.

In view of the aforementioned objective, we define the w*orst-case* m*akespan of a* t*emporal* n*etwork* (WCMTN) problem as follows.
INSTANCE. *A temporal network* $G = (V, E)$ *with* $V = \{1, \ldots, n\}$ *and* 1 *and* n *as unique source and sink, respectively. Vectors* $w, u \in \mathbb{N}_0^n$ *and scalars* $W, U \in \mathbb{N}_0$.
QUESTION. *Is there a* $\xi \in \Xi = \left\{\xi \in \mathbb{R}_+^n : \xi \leq \mathrm{e},\, w^\top \xi \leq W\right\}$ *such that*

$$\min_{y \in \mathbb{R}_+^n} \left\{y_n + u_n \xi_n \;:\; y_j \geq y_i + u_i \xi_i \quad \forall (i,j) \in E\right\} \;\geq\; U? \tag{6.3}$$

WCMTN considers instances of $\mathcal{RTN}$ with a fixed resource allocation $x \in X$, task durations that are linear in ξ and a support that results from intersecting the unit hypercube with a halfspace. WCMTN asks whether the worst-case makespan exceeds U when an optimal start time schedule is implemented.

Theorem 6.2.1 *WCMTN is* $\mathcal{NP}$*-complete.*

Proof. We first show that WCMTN belongs to $\mathcal{NP}$. Afterwards, we prove $\mathcal{NP}$-hardness of WCMTN by constructing a polynomial transformation of the Continuous Multiple Choice Knapsack problem to WCMTN. In this proof, we abbreviate "polynomial in the input length of WCMTN" by "polynomial".

To establish WCMTN's membership in $\mathcal{NP}$, we show that we can guess a ξ, check whether $\xi \in \Xi$, construct an admissible y^* that minimizes the left-hand side of the inequality (6.3) and verify whether $y_n^* + u_n \xi_n \geq U$ in polynomial time. Assume that we can restrict our attention to values of ξ whose bit lengths are polynomial. Then we can check in polynomial time whether $\xi \in \Xi$. Moreover, optimality of the early start schedule (see Sect. 6.2.1) ensures that y^* with $y_1^* = 0$ and $y_j^* = \max_{i \in V} \left\{y_i^* + u_i \xi_i : (i,j) \in E\right\}$ for $j \in V \setminus \{1\}$ minimizes the left-hand side of the inequality (6.3). In particular, this y^* also possesses a polynomial bit length and can be determined in polynomial time. This implies that the validity of the inequality (6.3) can be verified in polynomial time, which in turn implies membership of WCMTN in $\mathcal{NP}$. It remains to be shown that we can indeed restrict our attention to values of ξ with polynomial bit lengths. Note that the inequality (6.3) is satisfied for some $\xi \in \Xi$ if and only if

$$\max_{\xi \in \Xi} \min_{y \in \mathbb{R}_+^n} \left\{y_n + u_n \xi_n \;:\; y_j \geq y_i + u_i \xi_i \quad \forall (i,j) \in E\right\} \;\geq\; U.$$

Since the inner minimization is a convex function of ξ, its maximum over Ξ is attained by at least one extreme point of Ξ [HPT00]. Since Ξ is a polyhedron, however, all of its extreme points possess polynomial bit lengths [LP94].

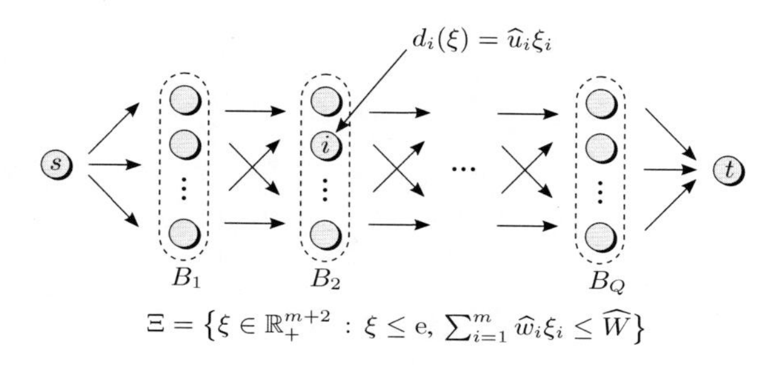

Fig. 6.2 WCMTN instance constructed from a CMCK instance

In order to prove $\mathcal{NP}$-hardness of WCMTN, we consider the Continuous Multiple Choice Knapsack (CMCK) problem [GJ79, Iba80]:
INSTANCE. *A set* $\mathcal{B} = \{1, \dots, m\}$*, together with weights* $\widehat{w}_i \in \mathbb{N}_0$ *and utilities* $\widehat{u}_i \in \mathbb{N}_0$ *for* $i \in \mathcal{B}$*. A partition* $\{B_q\}_{q=1}^Q$ *of* $\mathcal{B}$*, that is,* $\bigcup_q B_{q=1}^Q = \mathcal{B}$ *and* $B_q \cap B_r = \emptyset$ *for* $q \neq r$*. A maximum weight* $\widehat{W} \in \mathbb{N}_0$ *and a minimum utility* $\widehat{U} \in \mathbb{N}_0$.
QUESTION. *Is there a choice of* $b_q \in B_q$ *and* $\widehat{\xi}_q \in [0, 1]$*,* $q = 1, \dots, Q$*, such that* $\sum_{q=1}^Q \widehat{w}_{b_q}\widehat{\xi}_q \le \widehat{W}$ *and* $\sum_{q=1}^Q \widehat{u}_{b_q}\widehat{\xi}_q \ge \widehat{U}$*?*

We construct a polynomial-time transformation that converts a CMCK instance to a WCMTN instance such that the answer to the former problem is affirmative if and only if the answer to the latter one is.

The desired WCMTN instance is defined by $G = (V, E)$, $V = \{s, 1, \dots, m, t\}$ and $E = E_B \cup E_G$ with $E_B = \{(i, j) : (i, j) \in B_q \times B_{q+1}, q = 1, \dots, Q - 1\}$ and $E_G = \{(s, i) : i \in B_1\} \cup \{(i, t) : i \in B_Q\}$. The nodes s and t represent the unique source and sink of G, respectively. We set $w_i = \widehat{w}_i$ and $u_i = \widehat{u}_i$ for $i = 1, \dots, m$, while $w_i = u_i = 0$ for $i \in \{s, t\}$. We identify W and U with $\widehat{W}$ and $\widehat{U}$, respectively. The transformation is illustrated in Fig. 6.2.

For the constructed WCMTN instance, assume that there is a $\xi \in \Xi$ which satisfies the inequality (6.3). Let y^* be a minimizer for the left-hand side of (6.3). By construction of G and optimality of y^*, there is a critical path $(s, b_1, \dots, b_Q, t)$ in G with $b_q \in B_q$ for $q = 1, \dots, Q$, $y_s^* = y_{b_1}^* = 0$, $y_{b_{q+1}}^* = y_{b_q}^* + u_{b_q}\xi_{b_q}$ for $q = 1, \dots, Q - 1$ and $y_t^* = y_{b_Q}^* + u_{b_Q}\xi_{b_Q}$ [DH02]. Since $y_t^* \ge U$, we conclude that $\sum_{q=1}^Q u_{b_q}\xi_{b_q} = \sum_{q=1}^Q \widehat{u}_{b_q}\xi_{b_q} \ge U = \widehat{U}$. Similarly, we have $\sum_{q=1}^Q w_{b_q}\xi_{b_q} = \sum_{q=1}^Q \widehat{w}_{b_q}\xi_{b_q} \le W = \widehat{W}$ because $\xi \in \Xi$. Thus, b and $\widehat{\xi}$ with $\widehat{\xi}_q = \xi_{b_q}$, $q = 1, \dots, Q$, certify that the answer to the CMCK instance is affirmative as well. In the same way, one can show that the absence of a $\xi \in \Xi$ which satisfies the inequality (6.3) implies that the answer to the CMCK instance is negative. □

Theorem 6.2.1 extends to problem instances whose uncertainty sets are polyhedral [BS06] or that result from intersections of general ellipsoids as in [BTGN09]. However, it is easy to see that WCMTN can be decided in polynomial time for box uncertainty sets of the form $\Xi = \left\{\xi : \underline{\xi} \leq \xi \leq \overline{\xi}\right\}$ with $\underline{\xi}, \overline{\xi} \in \mathbb{R}^k$. The same holds true for the special case of WCMTN in which $w = \alpha\, \mathrm{e}$ and $u = \beta\, \mathrm{e}$ for $\alpha, \beta \in \mathbb{N}_0$.

We close with a review of two related complexity results. The complexity of optimization problems in temporal networks with probabilistic uncertainty is investigated in [Hag88]. In that paper the task durations are modeled as independent random variables with known, discrete distributions, and it is shown that calculating the mean or certain quantiles of the makespan distribution is #PSPACE-hard. We remark, however, that the worst-case duration (i.e., the 100%-quantile of the makespan distribution) can be calculated in polynomial time in that setting. In contrast, the additional complexity of WCMTN is due to the fact that our task durations are related through Ξ. The $\mathcal{NP}$-hardness of a generic robust resource allocation problem is proven in [KY97]. However, this problem is not defined on a network, and it assumes that X and Ξ are discrete.

6.3 Path-Wise Problem Formulation

In contrast to the techniques reviewed in Sect. 6.2.2, we will present a solution approach for $\mathcal{RTN}$ that does not approximate the optimal second-stage decision by decision rules. Instead, the approach eliminates the inner minimization in $\mathcal{RTN}$ by enumerating the task paths of the network. We have discussed a solution scheme based on path enumeration for two-stage chance constrained problems in the previous chapter. In this section, we present a path-wise reformulation of $\mathcal{RTN}$ and argue that its direct solution is prohibitive for temporal networks with large numbers of task paths. In the next two sections, we will use this path-wise reformulation to derive convergent bounds on the optimal value of $\mathcal{RTN}$.

We recall that a path in a directed graph $G = (V, E)$ constitutes a list of nodes $(i_1, \ldots, i_p)$ such that $(i_1, i_2), \ldots, (i_{p-1}, i_p) \in E$. Accordingly, we define a *task path* $P = \left\{i_1, \ldots, i_p\right\} \subseteq V$ as a set of tasks whose nodes form a path in the temporal network. We denote by $\mathcal{P}$ the set of all task paths. The following observation reiterates the well-known fact (see for example [DH02]) that for fixed x and ξ, the minimal makespan of a temporal network equals the sum of all task durations along any of its critical (i.e., most time-consuming) task paths.

Observation 6.3.1 *For a temporal network $G = (V, E)$ with fixed resource allocation $x \in X$ and parameters $\xi \in \Xi$, the minimal makespan is given by*

$$\min_{y \in Y(x,\xi)} \{y_n + d_n(x;\xi)\} = \max_{P \in \mathcal{P}} \left\{\mathbb{I}_P^\top d(x;\xi)\right\}, \tag{6.4}$$

where $d(x;\xi) = (d_1(x;\xi), \ldots, d_n(x;\xi))^\top$ and $Y(x,\xi)$ is defined in (6.1).

Note that the maximum on the right-hand side of the identity (6.4) can be attained by several task paths $P \in \mathcal{P}$. Observation 6.3.1 is crucial as it allows us to replace the inner minimization in $\mathcal{RTN}$ with a maximization. In analogy to

Observation 6.2.1, this reduces the two-stage robust optimization problem to an equivalent single-stage problem. Readers familiar with robust optimization may wonder whether a similar reduction can be achieved through duality arguments, see Sect. 2.2.2. Due to the structure of $Y(x, \xi)$, this approach results in a maximization problem whose objective function is nonconvex, and the resulting single-stage robust optimization problem would be difficult to solve. Observation 6.3.1 bypasses this problem at the expense of optimizing over a potentially large number of task paths.

Example 6.3.1. Consider the temporal network defined by the subgraph that contains the first four nodes in Fig. 6.1. Its minimal makespan is given by

$$\min_{y \in \mathbb{R}^4_+} \big\{ y_4 + d_4(x;\xi) \,:\, y_j \geq y_1 + d_1(x;\xi) \;\text{ for } j = 2, 3,$$

$$y_4 \geq y_j + d_j(x;\xi) \;\text{ for } j = 2, 3 \big\}.$$

By linear programming duality, this problem is equivalent to

$$\max_{\lambda \in \mathbb{R}^2_+} \big\{ [d_1(x;\xi) + d_2(x;\xi)]\, \lambda_1 + [d_1(x;\xi) + d_3(x;\xi)]\, \lambda_2 + d_4(x;\xi) \,:\, \lambda_1 + \lambda_2 \leq 1 \big\}.$$

For most task duration functions of interest, the objective function of this problem is nonconvex in ξ and λ. In contrast, enumerating the tasks paths yields

$$\max \left\{ d_1(x;\xi) + d_2(x;\xi) + d_4(x,\xi),\; d_1(x;\xi) + d_3(x;\xi) + d_4(x,\xi) \right\}.$$

The expressions in this maximization are convex in ξ if $d(x;\xi)$ is convex in ξ.

Applying Observation 6.3.1 to $\mathcal{RTN}$, we find

$$\min_{x \in X} \max_{\xi \in \Xi} \min_{y \in Y(x,\xi)} \{ y_n + d_n(x;\xi) \} \;=\; \min_{x \in X} \max_{P \in \mathcal{P}} \max_{\xi \in \Xi} \left\{ \mathbb{I}_P^\top d(x;\xi) \right\}.$$

In the following, we will employ robust optimization techniques to replace the maximization over Ξ. We are thus concerned with the following *approximate robust resource allocation problem on temporal networks*:

$$\min_{x \in X} \max_{P \in \mathcal{P}} \phi(x; P), \tag{$\mathcal{ARTN}$}$$

where $\phi(\cdot; P)$ represents a real-valued function on X. We call $\mathcal{ARTN}$ a *conservative reformulation* of $\mathcal{RTN}$ if

$$\phi(x; P) \geq \max_{\xi \in \Xi} \left\{ \mathbb{I}_P^\top d(x;\xi) \right\} \quad \text{for } x \in X,\; P \subseteq V. \tag{6.5}$$

If the inequality (6.5) is satisfied, then optimal allocations for $\mathcal{ARTN}$ constitute suboptimal but feasible allocations for $\mathcal{RTN}$, and the optimal value of $\mathcal{ARTN}$

overestimates the worst-case makespan in $\mathcal{RTN}$. If the inequality (6.5) can be replaced with an equality, then we call $\mathcal{ARTN}$ an *exact reformulation* of $\mathcal{RTN}$. In this case, $\mathcal{ARTN}$ and $\mathcal{RTN}$ are equivalent. The bounding approach presented in this chapter is applicable to exact and conservative reformulations of $\mathcal{RTN}$ alike. Note, however, that the method provides upper and lower bounds on $\mathcal{ARTN}$, and that these bounds will only bracket the optimal value of $\mathcal{RTN}$ if $\mathcal{ARTN}$ constitutes an exact reformulation.

Apart from $\mathcal{ARTN}$ being an exact or conservative reformulation of $\mathcal{RTN}$, the bounding approach requires ϕ to satisfy the following two properties:

(A1) *Monotonicity*: If $P \subset P' \subseteq V$, then $\phi(x; P) \leq \phi(x; P')$ for all $x \in X$.
(A2) *Sub-additivity*: If $P \subset P' \subseteq V$, then $\phi(x; P) + \phi(x; P' \setminus P) \geq \phi(x; P')$ for all $x \in X$.

We call $P \in \mathcal{P}$ an *inclusion-maximal path* if there is no $P' \in \mathcal{P}$, $P' \neq P$, such that $\mathbb{I}_P \leq \mathbb{I}_{P'}$. As in the previous chapter, we denote the set of inclusion-maximal paths by $\overline{\mathcal{P}} \subseteq \mathcal{P}$. If (A1) is satisfied, then the optimal allocations and the optimal value of $\mathcal{ARTN}$ do not change if we replace $\mathcal{P}$ with $\overline{\mathcal{P}}$. Property (A2) implies that $\phi(x; P)$ is bounded from above by $\sum_{r=1}^{R} \phi(x; P_r)$ for all $x \in X$ if $\{P_r\}_{r=1}^{R}$ forms a partition of P. As we will see, this bounding property facilitates the construction of lower and upper bounds on the optimal value of $\mathcal{ARTN}$. The following proposition shows that exact reformulations of $\mathcal{RTN}$ necessarily satisfy (A1) and (A2).

Proposition 6.3.1 *If $\mathcal{ARTN}$ is an exact reformulation of $\mathcal{RTN}$, then (A1) and (A2) are satisfied.*

Proof. For $P \subset P' \subseteq V$ and $x \in X$, we obtain

$$\phi(x; P') = \max_{\xi \in \Xi} \left\{ \mathbb{I}_{P'}^\top d(x; \xi) \right\} \geq \max_{\xi \in \Xi} \left\{ \mathbb{I}_P^\top d(x; \xi) \right\} = \phi(x; P),$$

where the inequality follows from $\mathbb{I}_{P'} \geq \mathbb{I}_P$ and nonnegativity of d. Similarly, for $P \subset P' \subseteq V$ and $x \in X$, we obtain

$$\begin{aligned} \phi(x; P') &= \max_{\xi \in \Xi} \left\{ \mathbb{I}_{P'}^\top d(x; \xi) \right\} \\ &= \max_{\xi \in \Xi} \left\{ \mathbb{I}_P^\top d(x; \xi) + \mathbb{I}_{[P' \setminus P]}^\top d(x; \xi) \right\} \\ &\leq \max_{\xi \in \Xi} \left\{ \mathbb{I}_P^\top d(x; \xi) \right\} + \max_{\xi \in \Xi} \left\{ \mathbb{I}_{[P' \setminus P]}^\top d(x; \xi) \right\} \\ &= \phi(x; P) + \phi(x; P' \setminus P). \end{aligned}$$

□

In the following, we focus on instances of $\mathcal{ARTN}$ that can be reformulated as explicit convex optimization problems. More precisely, we assume that

(A3) *Tractability*: X and $\phi(\cdot; P)$, $P \subseteq V$, possess tractable representations.

Remember that a set has a tractable representation if set membership can be described by finitely many convex constraints and auxiliary variables. Likewise, a function has a tractable representation if its epigraph does. Although the solution approach presented in this chapter does not rely on (A3), the repeated solution of lower and upper bound problems becomes computationally prohibitive if (A3) fails to hold. In the following, we show that robust optimization techniques allow us to construct exact or conservative reformulations of $\mathcal{RTN}$ that satisfy (A1)–(A3) for natural choices of X, Ξ and d.

Proposition 6.3.2 *If X has a tractable representation, then the following choices of Ξ and d allow for exact reformulations of $\mathcal{RTN}$ that satisfy (A1)–(A3):*

1. *Affine uncertainty: $d_i(x;\xi) = \delta_i^0(x) + \xi^\top[\delta_i^1(x)]$ with $\delta_i^0 : X \mapsto \mathbb{R}$ tractable, $\delta_i^1 : X \mapsto \mathbb{R}^k$ affine and $\xi \in \Xi = \bigcap_{l=1}^{L} \Xi_l \subseteq \mathbb{R}^k$ with*

$$\Xi_l = \left\{\xi \in \mathbb{R}^k \;:\; \exists u \in \mathbb{R}^{J_l} \text{ such that } \xi = \sigma^l + \Sigma^l u,\ \left\|\Pi^l u\right\|_2 \leq 1\right\},$$

   where $\sigma^l \in \mathbb{R}^k$, $\Sigma^l \in \mathbb{R}^{k\times J_l}$ and Π^l denotes a projection of $\mathbb{R}^{J_l}$ onto a subspace, $l = 1,\dots,L$. We require Ξ to be bounded and to have a nonempty relative interior.
2. *Quadratic uncertainty: $d_i(x;\xi) = \delta_i^0(x) + \xi^\top[\delta_i^1(x)] + \left\|[\Delta_i^2(x)]\,\xi\right\|_2^2$ with $\delta_i^0 : X \mapsto \mathbb{R}$ tractable, $\delta_i^1 : X \mapsto \mathbb{R}^k$ and $\Delta_i^2 : X \mapsto \mathbb{R}^{l\times k}$ affine and $\xi \in \Xi \subseteq \mathbb{R}^k$ with*

$$\Xi = \left\{\xi \in \mathbb{R}^k \;:\; \exists u \in \mathbb{R}^J \text{ such that } \xi = \sigma + \Sigma u,\ \|u\|_2 \leq 1\right\},$$

   where $\sigma \in \mathbb{R}^k$ and $\Sigma \in \mathbb{R}^{k\times J}$.

Proof. Let $\delta^0(x) = \left[\delta_1^0(x),\dots,\delta_n^0(x)\right]^\top$. In the case of affine uncertainty, we define ϕ through

$$\phi(x;P) = \mathbb{I}_P^\top[\delta^0(x)] + \max_{\xi\in\Xi}\left\{\xi^\top\Big(\sum_{i\in P}[\delta_i^1(x)]\Big)\right\} \quad \text{for } x \in X,\ P \in \mathcal{P},$$

and in the case of quadratic uncertainty, we define ϕ through

$$\phi(x;P) = \mathbb{I}_P^\top[\delta^0(x)] + \max_{\xi\in\Xi}\left\{\xi^\top\Big(\sum_{i\in P}[\delta_i^1(x)]\Big) + \left\|\operatorname{vec}\big([\mathbb{I}_P]_1\,[\Delta_1^2(x)]\xi,\dots,[\mathbb{I}_P]_n\,[\Delta_n^2(x)]\xi\big)\right\|_2^2\right\} \quad \text{for } x \in X,\ P \in \mathcal{P}.$$

Here, the operator "vec" returns the concatenation of its arguments as a column vector. We have $(\mathbb{I}_P)_i = 1$ if P contains task i and $(\mathbb{I}_P)_i = 0$ otherwise. Note that both definitions of $\phi(x;P)$ constitute exact reformulations of $\max_{\xi\in\Xi}\left\{\mathbb{I}_P^\top d(x;\xi)\right\}$. In either case, the epigraph of ϕ can be described by a semi-infinite constraint which has to hold for all $\xi \in \Xi$. Robust optimization techniques [BTGN09] enable us to reformulate these semi-infinite constraints such that (A3) is satisfied. Due to Proposition 6.3.1, (A1) and (A2) are satisfied as well. □

The uncertainty set considered in the first part of Proposition 6.3.2 covers all bounded polyhedra as special cases. Sometimes, the durations of the network tasks are best approximated by conic-quadratic functions, see Chap. 5. It is therefore desirable to extend the results of Proposition 6.3.2 also to problems with conic-quadratic uncertainty.

Proposition 6.3.3 (Conic-Quadratic Uncertainty) *Assume that the task durations are described by* $d_i(x;\xi) = \delta_i^0(x) + \xi^\top[\delta_i^1(x)] + \left\|[\Delta_i^2(x)]\,\xi\right\|_2$ *with* $\delta_i^0 : X \mapsto \mathbb{R}$ *tractable,* $\delta_i^1 : X \mapsto \mathbb{R}^k$ *and* $\Delta_i^2 : X \mapsto \mathbb{R}^{l\times k}$ *affine and* $\xi \in \Xi \subseteq \mathbb{R}^k$ *with*

$$\Xi = \left\{\xi \in \mathbb{R}^k \;:\; \exists u \in \mathbb{R}^J \;\; \text{such that} \;\; \xi = \sigma + \Sigma u,\; \|u\|_2 \leq 1\right\},$$

where $\sigma \in \mathbb{R}^k$, $\Sigma \in \mathbb{R}^{k\times J}$. *If* X *has a tractable representation, then* Ξ *and* d *allow for a conservative reformulation of* $\mathcal{RTN}$ *that satisfies (A1)–(A3).*

Remark 6.3.1. In contrast to the case of quadratic uncertainty, the last term of the task duration is not squared under conic-quadratic uncertainty.

Proof of Proposition 6.3.3 We construct an upper bound on

$$\max_{\xi\in\Xi}\left\{\sum_{i\in P}\left(\delta_i^0(x) + \xi^\top[\delta_i^1(x)] + \left\|[\Delta_i^2(x)]\xi\right\|_2\right)\right\} \quad \text{for } x \in X,\ P \in \mathcal{P}. \tag{6.6}$$

The terms in the objective of this problem either do not depend on ξ, or they are convex and linear homogeneous in ξ. Thus, we can apply the results from [BS06] and bound (6.6) from above by

$$\phi(x;P) = \max_{\widehat{u}\in\widehat{\mathcal{U}}}\left\{\mathbb{I}_P^\top[\widehat{d}(x;\widehat{u})]\right\}, \tag{6.7}$$

where $\widehat{u} = (\widehat{u}^+, \widehat{u}^-)$, and $\widehat{\mathcal{U}}$ is defined through

$$\widehat{\mathcal{U}} = \left\{\widehat{u} = (\widehat{u}^+, \widehat{u}^-) \in \mathbb{R}_+^J \times \mathbb{R}_+^J \;:\; \left\|\widehat{u}^+ + \widehat{u}^-\right\|_2 \leq 1\right\}.$$

Moreover, $\widehat{d} : X \times \mathbb{R}_+^{2J} \mapsto \mathbb{R}^n$ has components $\widehat{d}(x;\widehat{u}) = \left[\widehat{d}_1(x;\widehat{u}), \ldots, \widehat{d}_n(x;\widehat{u})\right]^\top$ that are defined through

$$\widehat{d}_i(x;\widehat{u}) = \delta_i^0(x) + \left[\sigma + \Sigma(\widehat{u}^+ - \widehat{u}^-)\right]^\top\left[\delta_i^1(x)\right] \\ + \underbrace{\left\|[\Delta_i^2(x)]\sigma\right\|_2 + \sum_{j=1}^J \left\|[\Delta_i^2(x)]\Sigma_j\right\|_2 (\widehat{u}_j^+ + \widehat{u}_j^-)}_{\alpha_i(x;\widehat{u})},$$

where Σ_j denotes the jth column of Σ. The epigraph of $\phi(x; P)$ can be described by a semi-infinite constraint that has to hold for all $\widehat{u} \in \widehat{\mathcal{U}}$. Due to the specific shape of $\widehat{\mathcal{U}}$ and the fact that $\widehat{d}$ is affine in $\widehat{u}$, robust optimization techniques can be employed to reformulate this semi-infinite constraint such that (A3) is satisfied. It remains to be shown that ϕ also satisfies (A1) and (A2).

As for (A1), we show that $\widehat{d}_i(x;\widehat{u})$, $i \in V$, is nonnegative for all $x \in X$ and $\widehat{u} \in \widehat{\mathcal{U}}$. To this end, we fix some $\widehat{u} = (\widehat{u}^+, \widehat{u}^-) \in \widehat{\mathcal{U}}$ and set $u = \widehat{u}^+ - \widehat{u}^-$. Then $\xi = \sigma + \Sigma u$ is contained in Ξ since $\|u\|_2 \leq 1$. Hence, for $x \in X$ we have

$$d_i(x;\xi) = \delta_i^0(x) + \left[\sigma + \Sigma u\right]^\top [\delta_i^1(x)] + \underbrace{\left\|[\Delta_i^2(x)][\sigma + \Sigma u]\right\|_2}_{\beta_i(x;u)} \geq 0$$

by nonnegativity of d. Note that $\widehat{d}_i(x;\widehat{u}) - d_i(x;\xi) = \alpha_i(x;\widehat{u}) - \beta_i(x;u)$ for this choice of ξ. Since $d_i(x;\xi) \geq 0$, nonnegativity of $\widehat{d}_i(x;\widehat{u})$ is ensured if $\alpha_i(x;\widehat{u}) \geq \beta_i(x;u)$. The latter inequality follows from the triangle inequality, the positive homogeneity of norms and the fact that $|u_j| \leq \widehat{u}_j^+ + \widehat{u}_j^-$.

As for (A2), we need to show that $\phi(x; P) + \phi(x; P' \setminus P) \geq \phi(x; P')$ for $x \in X$ and $P \subset P' \subseteq V$. This is the case since

$$\max_{\widehat{u}\in\widehat{\mathcal{U}}} \left\{\mathbb{I}_P^\top\left[\widehat{d}(x;\widehat{u})\right]\right\} + \max_{\widehat{u}\in\widehat{\mathcal{U}}} \left\{\mathbb{I}_{[P'\setminus P]}^\top\left[\widehat{d}(x;\widehat{u})\right]\right\} \geq \max_{\widehat{u}\in\widehat{\mathcal{U}}} \left\{\mathbb{I}_{P'}^\top\left[\widehat{d}(x;\widehat{u})\right]\right\}.$$

□

Proposition 6.3.3 provides a conservative reformulation of $\mathcal{RTN}$. Exact reformulations of robust optimization problems subject to conic-quadratic uncertainty are discussed in [BTGN09]. However, the path durations $\phi(x; P)$ resulting from conic-quadratic uncertainty are not of the form required in [BTGN09], and the corresponding reformulation does not seem to be applicable to our context.

Note that even if (A3) is satisfied, $\mathcal{ARTN}$ remains generically intractable since its size grows with the cardinality of $\mathcal{P}$, which in turn can be exponential in the size of G. Indeed, one can show that the expected number of paths in a uniformly sampled random temporal network is exponential.

Theorem 6.3.1 *For a fixed connectivity $\rho \in (0, 1]$ and network size $n \in \mathbb{N}$, consider a random temporal network $G = (V, E)$ with $V = \{1, \dots, n\}$ that is constructed as follows. For each node $i \in V \setminus \{n\}$, we choose the number of immediate successors $\{1, \dots, \lceil\rho(n-i)\rceil\}$ uniformly at random. Afterwards, we choose the indices of the successor nodes from $\{i+1, \dots, n\}$, again uniformly at random. Then, the expected number of paths in G is exponential in n.*

Proof. By construction, G is acyclic and has the unique sink n. The probability that j is a successor of i, $i < j$, is

$$\frac{1}{\lceil\rho(n-i)\rceil} \sum_{j=1}^{\lceil\rho(n-i)\rceil} \frac{j}{n-i} = \frac{\lceil\rho(n-i)\rceil\left(\lceil\rho(n-i)\rceil + 1\right)}{2\lceil\rho(n-i)\rceil(n-i)} = \frac{\lceil\rho(n-i)\rceil + 1}{2(n-i)}.$$

Let X_i be the random variable that describes the number of paths from node i to node n. We have $\mathbb{E}(X_n) = 1$ and obtain

$$\mathbb{E}(X_i) = \frac{\lceil \rho(n-i) \rceil + 1}{2(n-i)} \sum_{j=i+1}^{n} \mathbb{E}(X_j) \quad \text{for } i < n.$$

In particular, $\mathbb{E}(X_{n-1}) = 1$. For $i < n$, we can express $\mathbb{E}(X_i)$ as follows:

$$\begin{aligned}
\mathbb{E}(X_i) &= \frac{\lceil \rho(n-i) \rceil + 1}{2(n-i)} \left(1 + \frac{2(n-i-1)}{\lceil \rho(n-i-1) \rceil + 1}\right) \mathbb{E}(X_{i+1}) \\
&= \frac{\lceil \rho(n-i) \rceil + 1}{2(n-i)} \frac{\lceil \rho(n-i-1) \rceil + 1 + 2(n-i-1)}{\lceil \rho(n-i-1) \rceil + 1} \mathbb{E}(X_{i+1}).
\end{aligned}$$

Partially unrolling the recursion, we obtain for $\mathbb{E}(X_1)$ and $m \in \{2, \ldots, n\}$:

$$\begin{aligned}
\mathbb{E}(X_1) &= \left(\prod_{i=1}^{m-1} \frac{\lceil \rho(n-i) \rceil + 1}{2(n-i)} \frac{\lceil \rho(n-i-1) \rceil + 1 + 2(n-i-1)}{\lceil \rho(n-i-1) \rceil + 1} \right) \mathbb{E}(X_m) \\
&= \frac{\lceil \rho(n-1) \rceil + 1}{\lceil \rho(n-m) \rceil + 1} \left(\prod_{i=1}^{m-1} \frac{\lceil \rho(n-i-1) \rceil + 1 + 2(n-i-1)}{2(n-i)} \right) \mathbb{E}(X_m) \\
&= \frac{\lceil \rho(n-1) \rceil + 1}{\lceil \rho(n-m) \rceil + 1} \left(\prod_{i=1}^{m-1} \left[1 + \frac{\lceil \rho(n-i-1) \rceil - 1}{2(n-i)} \right] \right) \mathbb{E}(X_m).
\end{aligned}$$

Let us investigate the term $(\lceil \rho(n-i-1) \rceil - 1)/(2[n-i])$. We show that for a specific choice of m, this term is greater than or equal to some $\delta > 0$. Note that

$$\frac{\lceil \rho(n-i-1) \rceil - 1}{2(n-i)} \geq \frac{\rho(n-i-1) - 1}{2(n-i)} = \frac{\rho(n-i) - \rho - 1}{2(n-i)} = \frac{\rho}{2} - \frac{\rho + 1}{2(n-i)}.$$

Assume that $n \geq 2/\rho + 4$. Then the last expression is greater than or equal to $\rho/4$, a strictly positive number, for all $i \leq \overline{m} = n - \lceil (2\rho + 2)/\rho \rceil$. We obtain

$$\begin{aligned}
\mathbb{E}(X_1) &= \frac{\lceil \rho(n-1) \rceil + 1}{\lceil \rho(n-\overline{m}) \rceil + 1} \left(\prod_{i=1}^{\overline{m}-1} \left(1 + \frac{\lceil \rho(n-i-1) \rceil - 1}{2(n-i)} \right) \right) \mathbb{E}(X_{\overline{m}}) \\
&\geq \frac{\lceil \rho(n-1) \rceil + 1}{\lceil \rho(n-\overline{m}) \rceil + 1} \prod_{i=1}^{\overline{m}-1} \left(1 + \frac{\lceil \rho(n-i-1) \rceil - 1}{2(n-i)} \right) \\
&\geq \frac{\lceil \rho(n-1) \rceil + 1}{\lceil \rho(n-\overline{m}) \rceil + 1} \prod_{i=1}^{\overline{m}-1} \left(1 + \frac{\rho}{4} \right)
\end{aligned}$$

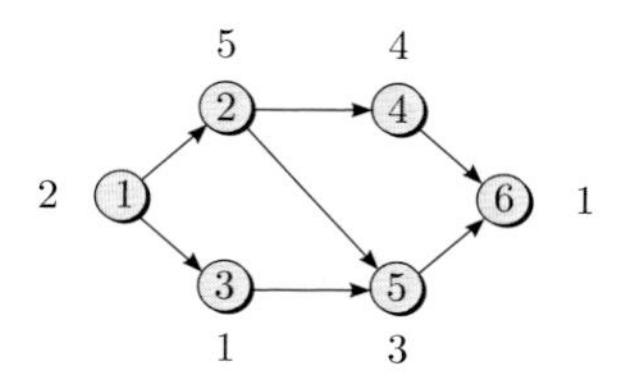

Fig. 6.3 Example temporal network. The chart illustrates the network structure and the nominal durations d_i^0 of the network tasks $i \in V$ (attached to the nodes)

$$= \frac{\lceil \rho(n-1) \rceil + 1}{\lceil \rho(n-\overline{m}) \rceil + 1} \left(1 + \frac{\rho}{4}\right)^{\overline{m}-1} \in \Omega\left(n(1+\rho/4)^n\right),$$

where $\Omega(\cdot)$ denotes the asymptotic lower bound in Bachmann–Landau notation. Since the expected number of paths from node 1 to node n is already exponential, the expected number of all paths in network G is exponential, too. □

Hence, even though $\mathcal{ARTN}$ can be expressed as an explicit convex optimization problem, it remains difficult to solve. We close with an example that illustrates the path-wise problem formulation $\mathcal{ARTN}$.

Example 6.3.2. Consider the temporal network in Fig. 6.3. Apart from the missing cash flows, it is identical to the temporal network in Fig. 1.1. Now, however, we interpret the number attached to task $i \in V$ as the nominal duration of task i. We consider a resource allocation problem with one resource and task durations

$$d_i(x;\xi) = d_i^0\,(1-x_i)\,(1+\xi_i) \qquad \text{for } i \in V,$$

where d_i^0 denotes the nominal task duration from Fig. 6.3, x_i the amount of the resource that is assigned to task i, and ξ_i the uncertainty inherent to the task duration. We set

$$X = \left\{x \in \mathbb{R}^6_+ \,:\, x_i \leq 1/2,\ \mathrm{e}^\top x \leq 1\right\}$$
$$\text{and} \quad \Xi = \left\{\xi \in \mathbb{R}^6_+ \,:\, \xi_i \leq 1/2,\ \mathrm{e}^\top \xi \leq 1\right\}.$$

Thus, the duration of task i can fall below or exceed its nominal duration d_i^0 by 50%, depending on the resource allocation and the realization of the uncertain parameter vector ξ. Up to two tasks can be sped up to their minimal durations, and up to two tasks on each inclusion-maximal path can attain their worst-case durations.

For the network in Fig. 6.3, the set $\mathcal{P}$ of all task paths contains 22 elements. Elements of $\mathcal{P}$ are, amongst others, $\{1\}, \{2\}, \ldots, \{6\}, \{1,2\}, \{1,3\}, \ldots, \{5,6\}$ and $\{1,2,5\}$. The set $\overline{\mathcal{P}}$ of inclusion-maximal task paths only contains three elements, namely $\{1,2,4,6\}$, $\{1,2,5,6\}$ and $\{1,3,5,6\}$. Since the problem instance satisfies the conditions of the first part of Proposition 6.3.2, we can develop an exact reformulation of $\mathcal{RTN}$ that satisfies (A1)–(A3). Indeed, for our choice of functions we have

$$\phi(x;P) = \max_{\xi_i \in \mathbb{R}_+ : i \in P} \left\{ \sum_{i \in P} d_i^0(1-x_i)(1+\xi_i) : \xi_i \leq 1/2 \quad \forall i \in P, \quad \sum_{i \in P} \xi_i \leq 1 \right\}$$

$$= \min_{\substack{\lambda_i \in \mathbb{R}_+ : i \in P, \\ \gamma \in \mathbb{R}_+}} \left\{ \left[\sum_{i \in P} d_i^0(1-x_i) + \lambda_i/2 \right] + \gamma : \lambda_i + \gamma \geq d_i^0(1-x_i) \quad \forall i \in P \right\},$$

where the first identity holds by definition, and the second one follows from linear programming duality. Note that our reformulation $\mathcal{ARTN}$ satisfies (A3) since

$$\tau \geq \phi(x;P) \Leftrightarrow \exists (\lambda_i \in \mathbb{R}_+ : i \in P), \gamma \in \mathbb{R}_+ : \tau \geq \left[\sum_{i \in P} d_i^0(1-x_i) + \lambda_i/2 \right] + \gamma,$$

$$\lambda_i + \gamma \geq d_i^0(1-x_i) \quad \forall i \in P,$$

and the right-hand side of this equivalence can be expressed by finitely many linear constraints and auxiliary variables. For the temporal network in Fig. 6.3, our reformulation $\mathcal{ARTN}$ results in the following optimization problem.

$$\begin{aligned}
\underset{\tau,x,\lambda,\gamma}{\text{minimize}} \quad & \tau \\
\text{subject to} \quad & \tau \in \mathbb{R}_+, \quad x \in \mathbb{R}_+^6, \quad \lambda \in \mathbb{R}_+^{12}, \quad \gamma \in \mathbb{R}_+^3 \\
& \tau \geq 2(1-x_1) + \lambda_1^1/2 + 5(1-x_2) + \lambda_2^1/2 \\
& \qquad + 4(1-x_4) + \lambda_4^1/2 + 1(1-x_6) + \lambda_6^1/2 + \gamma^1, \\
& \lambda_1^1 + \gamma^1 \geq 2(1-x_1), \quad \lambda_2^1 + \gamma^1 \geq 5(1-x_2), \\
& \lambda_4^1 + \gamma^1 \geq 4(1-x_4), \quad \lambda_6^1 + \gamma^1 \geq 1(1-x_6), \\
& \tau \geq 2(1-x_1) + \lambda_1^2/2 + 5(1-x_2) + \lambda_2^2/2 \\
& \qquad + 3(1-x_5) + \lambda_5^2/2 + 1(1-x_6) + \lambda_6^2/2 + \gamma^2, \\
& \lambda_1^2 + \gamma^2 \geq 2(1-x_1), \quad \lambda_2^2 + \gamma^2 \geq 5(1-x_2), \\
& \lambda_5^2 + \gamma^2 \geq 3(1-x_5), \quad \lambda_6^2 + \gamma^2 \geq 1(1-x_6), \\
& \tau \geq 2(1-x_1) + \lambda_1^3/2 + 1(1-x_3) + \lambda_3^3/2 \\
& \qquad + 3(1-x_5) + \lambda_5^3/2 + 1(1-x_6) + \lambda_6^3/2 + \gamma^3, \\
& \lambda_1^3 + \gamma^3 \geq 2(1-x_1), \quad \lambda_3^3 + \gamma^3 \geq 1(1-x_3),
\end{aligned}$$

$$\lambda_5^3 + \gamma^3 \geq 3(1 - x_5), \quad \lambda_6^3 + \gamma^3 \geq 1(1 - x_6),$$

$$x_i \leq 1/2 \;\; \forall i \in \{1, \ldots, 6\}, \quad \sum_{i=1}^{6} x_i \leq 1.$$

The optimal allocation for this problem is $x = (0, 0.50, 0, 0.36, 0.14, 0)^\top$ and leads to a worst-case makespan of 10.61.

6.4 Lower Bounds

The convergent lower bounds on $\mathcal{ARTN}$ are obtained by solving relaxations that omit some of the paths in $\mathcal{ARTN}$. The procedure is described in Algorithm 4.

Algorithm 4 Convergent lower bounds on $\mathcal{ARTN}$.

1. **Initialization.** Choose a subset $\mathcal{P}_1 \subseteq \overline{\mathcal{P}}$, for example $\mathcal{P}_1 = \emptyset$. Set $t = 1$.
2. **Master Problem.** Solve $\mathcal{ARTN}$, restricted to the paths in $\mathcal{P}_t$:

$$\begin{aligned} \underset{x,\tau}{\text{minimize}} \quad & \tau \\ \text{subject to} \quad & x \in X, \;\; \tau \in \mathbb{R}_+ \\ & \tau \geq \phi(x; P) \qquad \forall P \in \mathcal{P}_t. \end{aligned} \qquad (\mathcal{LARTN}_t)$$

   Let x^t denote an optimal solution to $\mathcal{LARTN}_t$ and τ^t its objective value.
3. **Subproblem.** Determine a path $P \in \overline{\mathcal{P}} \setminus \mathcal{P}_t$ with $\phi(x^t; P) > \tau^t$.

   a. *If no such path exists*, stop: $x^* = x^t$ constitutes an optimal solution to $\mathcal{ARTN}$ and $\tau^* = \tau^t$ its objective value.
   b. *Otherwise*, set $\mathcal{P}_{t+1} = \mathcal{P}_t \cup \{P\}$, $t \to t + 1$ and go to Step 2.

The following proposition is an immediate consequence of the algorithm outline.

Proposition 6.4.1 *Algorithm 4 terminates with an optimal allocation x^* for $\mathcal{ARTN}$, together with its worst-case makespan τ^*. Furthermore, $\{\tau^t\}_t$ represents a monotonically non-decreasing sequence of lower bounds on τ^*.*

Proof. Since $t \leq t'$ implies that $\mathcal{P}_t \subseteq \mathcal{P}_{t'}$, $\mathcal{LARTN}_t$ constitutes a relaxation of $\mathcal{LARTN}_{t'}$. Hence, $\tau^t \leq \tau^{t'}$, that is, $\{\tau^t\}_t$ is monotonically non-decreasing. Similarly, every τ^t constitutes a lower bound on the optimal value of $\mathcal{ARTN}$, because the latter problem considers all paths in $\overline{\mathcal{P}}$ and $\mathcal{P}_t \subseteq \overline{\mathcal{P}}$ for all t.

In iteration t, Step 3 either terminates or adds a path $P \in \overline{\mathcal{P}} \setminus \mathcal{P}_t$ to $\mathcal{P}_t$. Hence, the algorithm terminates after $T \leq \left|\overline{\mathcal{P}} \setminus \mathcal{P}_1\right| + 1$ iterations. It is clear that x^* is optimal if $\mathcal{P}_T = \overline{\mathcal{P}}$ in the last iteration. Otherwise, $\phi(x^*; P) \leq \tau^*$ for all $P \in \overline{\mathcal{P}} \setminus \mathcal{P}_T$. Thus, (x^*, τ^*) minimizes the relaxation $\mathcal{LARTN}_T$ and x^* is feasible in $\mathcal{ARTN}$.

Since x^* attains the same objective value τ^* in $\mathcal{ARTN}$, x^* is an optimal allocation and τ^* the optimal value of $\mathcal{ARTN}$. □

The size of $\mathcal{LARTN}_t$, $t \geq 1$, grows with the cardinality of $\mathcal{P}_t$. Hence, Algorithm 4 allows us to determine coarse initial lower bounds with little effort, whereas tighter lower bounds become increasingly difficult to obtain.

The quality of the lower bounds determined by Algorithm 4 crucially depends on the path selection in Step 3. In iteration t it seems natural to select a path P that maximizes $\phi(x^t; P)$ over $\overline{\mathcal{P}} \setminus \mathcal{P}_t$. Theorem 6.2.1 implies that this choice may require the solution of an $\mathcal{NP}$-hard optimization problem. A naive alternative is to enumerate all paths in $\overline{\mathcal{P}} \setminus \mathcal{P}_t$ and stop once a path P is found that satisfies $\phi(x^t; P) > \tau^t$. This "first fit" method, however, suffers from two limitations. Firstly, this approach is likely to require many iterations since there is no prioritization among the paths P that satisfy $\phi(x^t; P) > \tau^t$. Secondly, in the last (Tth) iteration of Algorithm 4 all paths in $\overline{\mathcal{P}} \setminus \mathcal{P}_T$ are investigated before the procedure can terminate. This implies that the algorithm needs to inspect all elements of $\overline{\mathcal{P}}$ at least once. In view of the cardinality of $\overline{\mathcal{P}}$ (see Sect. 6.3), this is computationally prohibitive. To alleviate both problems, we replace Step 3 of Algorithm 4 with Algorithm 5.

Algorithm 5 Determine $P \in \overline{\mathcal{P}} \setminus \mathcal{P}_t$ with $\phi(x^t; P) > \tau^t$.

3(a) **Initialization.** Construct the temporal network $G = (V, E)$ with deterministic task durations $\delta = (\delta_1, \ldots, \delta_n)^\top$, where $\delta_i = \max\{\phi(x^t; \{i\}), \epsilon\}$. Here, $\{i\}$ represents a degenerate path that contains a single task $i \in V$, while ϵ denotes a small positive constant. Set $s = 1$.

3(b) **Path Selection.** Let P_s be the sth longest path in G, where the length of a path $P \in \mathcal{P}$ is defined as $\mathbb{I}_P^\top \delta$.

(i) *If* $\mathbb{I}_{P_s}^\top \delta \leq \tau^t$ *or G contains less than s paths*, stop: $x^* = x^t$ is an optimal allocation in $\mathcal{ARTN}$ and $\tau^* = \tau^t$ its worst-case makespan.

(ii) *If* $\phi(x^t; P_s) > \tau^t$, set $\mathcal{P}_{t+1} = \mathcal{P}_t \cup \{P_s\}$, $t \to t + 1$ and go to Step 2 of Algorithm 4.

(iii) *Otherwise*, set $s \to s + 1$ and repeat Step 3(b).

The algorithm uses $\mathbb{I}_P^\top \delta$ as an overestimator for $\phi(x^t; P)$. Indeed, we have $\mathbb{I}_P^\top \delta \geq \sum_{i \in P} \phi(x^t; \{i\})$ by definition of δ, while $\sum_{i \in P} \phi(x^t; \{i\})$ exceeds $\phi(x^t; P)$ due to (A2). Note that $\phi(x^t; \{i\})$ represents the worst-case duration of task i.

Depending on the problem instance, Algorithm 5 may certify the optimality of x^t without inspecting all paths in $\mathcal{P}$. Furthermore, if ϵ is sufficiently small, then the paths $P \in \mathcal{P}$ are inspected in the order of decreasing task-wise worst-case durations $\sum_{i \in P} \phi(x^t; \{i\})$. Thus, as long as these quantities approximate $\phi(x^t; P)$, $P \in \mathcal{P}$, reasonably well, one can expect Algorithm 5 to outperform the "first fit" approach outlined above. Note that the s longest paths in a directed, acyclic graph $G = (V, E)$ can be enumerated in time $\mathcal{O}(|E| + s\,|V|)$, see [Epp94]. The following proposition establishes the correctness of Algorithm 5.

Proposition 6.4.2 *Algorithm 5 terminates and either correctly concludes that x^t is an optimal allocation in $\mathcal{ARTN}$, or it determines a path $P \in \overline{\mathcal{P}} \setminus \mathcal{P}_t$ with $\phi(x^t; P) > \tau^t$.*

Proof. G contains a finite number of paths, and hence the algorithm terminates. In the following, we denote by P_s the sth longest path in G according to the metric defined in Step 3(b) of the algorithm. Furthermore, we assume that the algorithm terminates in iteration S.

Assume that the algorithm terminates in case (i) of Step 3(b) because G contains less than S paths. In this case, all paths $P \in \mathcal{P}$ satisfy $\tau^t \geq \phi(x^t; P)$ since otherwise the algorithm would have terminated in case (ii) of Step 3(b) of an earlier iteration. From Proposition 6.4.1 we conclude that x^t constitutes an optimal allocation in $\mathcal{ARTN}$.

If the algorithm terminates in case (i) of Step 3(b) because $\mathbb{I}_{P_S}^\top \delta \leq \tau^t$, then we know that $\tau^t \geq \phi(x^t; P_s)$ for all $s < S$. Also, $\tau^t \geq \mathbb{I}_{P_s}^\top \delta$ for $s \in \{S+1, \ldots, |\mathcal{P}|\}$ since these paths are not longer than P_S. This, however, implies that for $P \in \{P_S, \ldots, P_{|\mathcal{P}|}\}$, we have

$$\tau^t \;\geq\; \mathbb{I}_P^\top \delta \;\geq\; \sum_{i \in P} \phi(x^t; \{i\}) \;\geq\; \phi(x^t; P),$$

where δ is defined in Step 3(a) of Algorithm 5. The second inequality follows from the definition of δ, while the third one is due to (A2). We conclude that $\tau^t \geq \phi(x^t; P)$ for all $P \in \mathcal{P}$, and hence Proposition 6.4.1 ensures that x^t is an optimal allocation in $\mathcal{ARTN}$.

If the algorithm terminates in case (ii) of Step 3(b), then it has determined a task path $P_S \in \mathcal{P}$ with $\phi(x^t; P_S) > \tau^t$. We need to show that P_S is inclusion-maximal, that is, $P_S \in \overline{\mathcal{P}}$. Assume to the contrary that $P_S \in \mathcal{P} \setminus \overline{\mathcal{P}}$. Then there is a task path $P \in \mathcal{P}$ with $P \neq P_S$ and $\mathbb{I}_P \geq \mathbb{I}_{P_S}$. Since $\delta > 0$ component-wise, we have $\mathbb{I}_P^\top \delta = \left(\mathbb{I}_{P_S} + \mathbb{I}_{[P \setminus P_S]}\right)^\top \delta > \mathbb{I}_{P_S}^\top \delta$. Hence, P must have been considered in some iteration $s < S$. Due to (A1), however, we have $\phi(x^t; P) \geq \phi(x^t; P_S)$, and the algorithm must have terminated in case (ii) of Step 3(b) of that iteration because $\phi(x^t; P) \geq \phi(x^t; P_S) > \tau^t$. Since this yields a contradiction, we conclude that P_S is indeed inclusion-maximal. □

Note that prior to its termination, Algorithm 4 only provides monotonically increasing *lower* bounds on the optimal value of $\mathcal{ARTN}$. Since the intermediate allocations x^t are feasible, their worst-case makespans in $\mathcal{ARTN}$ also constitute *upper* bounds on the optimal value of $\mathcal{ARTN}$. From Theorem 6.2.1, however, we know that evaluating the worst-case makespan of x^t in $\mathcal{ARTN}$ may require the solution of an $\mathcal{NP}$-hard optimization problem. Hence, we need to pursue a different approach to generate upper bounds efficiently.

We close with an example that illustrates Algorithms 4 and 5.

Example 6.4.1. Consider again the resource allocation problem defined in Example 6.2.1. We generate lower bounds on the optimal objective value of this problem with Algorithms 4 and 5.

We start with *Step 1* of Algorithm 4, in which we choose the subset $\mathcal{P}_1 = \emptyset$ and set $t = 1$.

In *Step 2* we solve the following lower bound problem $\mathcal{LARTN}_1$:

$$\begin{aligned} \underset{\tau,x}{\text{minimize}} \quad & \tau \\ \text{subject to} \quad & \tau \in \mathbb{R}_+, \quad x \in \mathbb{R}^6_+ \\ & x_i \leq 1/2 \quad \forall i \in \{1,\ldots,6\}, \quad \sum_{i=1}^{6} x_i \leq 1. \end{aligned}$$

The optimal allocation is $x^1 = (0,0,0,0,0,0)^\top$ with an estimated worst-case makespan of $\tau^1 = 0$.

We now enter *Step 3(a)* of Algorithm 5. The deterministic temporal network with worst-case task durations $\delta = (3, 7.5, 1.5, 6, 4.5, 1.5)^\top$ is illustrated in Fig 6.4, upper left.

In *Step 3(b)*, we identify $P_1 = \{1, 2, 4, 6\}$ as the longest path in the deterministic temporal network. This path has a task-wise worst-case duration of $\mathbb{I}_{P_1}^\top \delta = 18$ and a path-wise worst-case duration of $\phi(x^1; P_1) = 16.5$. This path therefore satisfies condition (ii) of Step 3(b), and we set $\mathcal{P}_2 = \{\{1, 2, 4, 6\}\}$ and $t = 2$.

We are back in *Step 2* of Algorithm 4. The new lower bound $\mathcal{LARTN}_2$ is obtained from the following optimization problem:

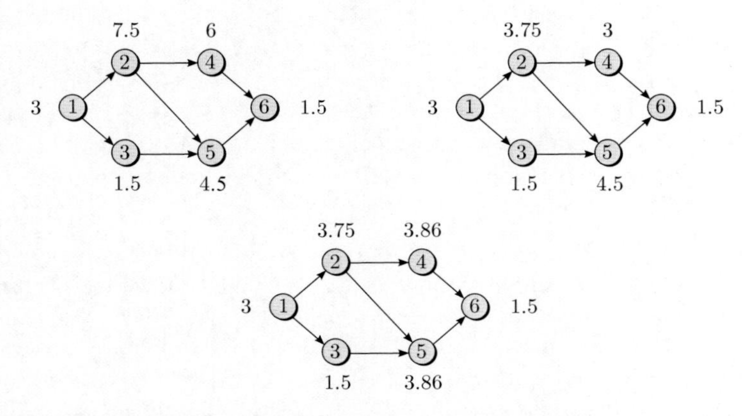

Fig. 6.4 Auxiliary deterministic temporal networks generated by Algorithm 5. The *upper left, upper right and lower charts* visualize the auxiliary graphs in iteration $t = 1$, $t = 2$ and $t = 3$, respectively. Attached to each node $i \in V$ is its task-wise worst-case duration δ_i

$$
\begin{aligned}
\underset{\tau,x,\lambda,\gamma}{\text{minimize}} \quad & \tau \\
\text{subject to} \quad & \tau \in \mathbb{R}_+, \quad x \in \mathbb{R}^6_+, \quad \lambda \in \mathbb{R}^4_+, \quad \gamma \in \mathbb{R}_+ \\
& \tau \geq 2(1-x_1) + \lambda^1_1/2 + 5(1-x_2) + \lambda^1_2/2 \\
& \qquad + 4(1-x_4) + \lambda^1_4/2 + 1(1-x_6) + \lambda^1_6/2 + \gamma^1, \\
& \lambda^1_1 + \gamma^1 \geq 2(1-x_1), \quad \lambda^1_2 + \gamma^1 \geq 5(1-x_2), \\
& \lambda^1_4 + \gamma^1 \geq 4(1-x_4), \quad \lambda^1_6 + \gamma^1 \geq 1(1-x_6), \\
& x_i \leq 1/2 \;\; \forall i \in \{1,\ldots,6\}, \quad \sum_{i=1}^{6} x_i \leq 1.
\end{aligned}
$$

The optimal allocation to this problem is $x^2 = (0, 0.5, 0, 0.5, 0, 0)^\top$ and leads to an estimated worst-case makespan of $\tau^2 = 9.75$.

We enter *Step* 3(*a*) of Algorithm 5 again. Figure 6.4, upper right illustrates the deterministic temporal network with worst-case task durations $\delta = (3, 3.75, 1.5, 3, 4.5, 1.5)^\top$. We set $s = 1$.

In *Step* 3(*b*), we identify $P_1 = \{1, 2, 5, 6\}$ as the longest path in the deterministic temporal network. This path has a task-wise worst-case duration of $\mathbb{I}^\top_{P_1}\delta = 12.75$ and a path-wise worst-case duration of $\phi(x^2; P_1) = 11.25$. This path therefore satisfies condition (ii) of Step 3(b), and we set $\mathcal{P}_3 = \{\{1, 2, 4, 6\}, \{1, 2, 5, 6\}\}$ and $t = 3$.

We are back in *Step 2* of Algorithm 4. The new lower bound $\mathcal{LARTN}_3$ is obtained from the following optimization problem:

$$
\begin{aligned}
\underset{\tau,x,\lambda,\gamma}{\text{minimize}} \quad & \tau \\
\text{subject to} \quad & \tau \in \mathbb{R}_+, \quad x \in \mathbb{R}^6_+, \quad \lambda \in \mathbb{R}^8_+, \quad \gamma \in \mathbb{R}^2_+ \\
& \tau \geq 2(1-x_1) + \lambda^1_1/2 + 5(1-x_2) + \lambda^1_2/2 \\
& \qquad + 4(1-x_4) + \lambda^1_4/2 + 1(1-x_6) + \lambda^1_6/2 + \gamma^1, \\
& \lambda^1_1 + \gamma^1 \geq 2(1-x_1), \quad \lambda^1_2 + \gamma^1 \geq 5(1-x_2), \\
& \lambda^1_4 + \gamma^1 \geq 4(1-x_4), \quad \lambda^1_6 + \gamma^1 \geq 1(1-x_6), \\
& \tau \geq 2(1-x_1) + \lambda^2_1/2 + 5(1-x_2) + \lambda^2_2/2 \\
& \qquad + 3(1-x_5) + \lambda^2_5/2 + 1(1-x_6) + \lambda^2_6/2 + \gamma^2, \\
& \lambda^2_1 + \gamma^2 \geq 2(1-x_1), \quad \lambda^2_2 + \gamma^2 \geq 5(1-x_2),
\end{aligned}
$$

$$\lambda_5^2 + \gamma^2 \geq 3(1 - x_5), \quad \lambda_6^2 + \gamma^2 \geq 1(1 - x_6),$$

$$x_i \leq 1/2 \ \ \forall i \in \{1, \ldots, 6\}, \quad \sum_{i=1}^{6} x_i \leq 1.$$

The optimal allocation to this problem is $x^3 = (0, 0.5, 0, 0.36, 0.14, 0)^\top$ and leads to an estimated worst-case makespan of $\tau^3 = 10.61$.

We enter $Step\ 3(a)$ of Algorithm 5 again. The deterministic temporal network with worst-case task durations $\delta = (3, 3.75, 1.5, 3.86, 3.86, 1.5)^\top$ is illustrated in Fig. 6.4, lower chart. We set $s = 1$.

In $Step\ 3(b)$, we identify $P_1 = \{1, 2, 4, 6\}$ as the longest path in the deterministic temporal network. This path has a task-wise worst-case duration of $\mathbb{I}_{P_1}^\top \delta = 12.11$ and a path-wise worst-case duration of $\phi(x^3; P_1) = 10.61$. This path therefore satisfies condition (iii) of Step 3(b), and we set $s = 2$.

In $Step\ 3(b)$, we identify $P_2 = \{1, 2, 5, 6\}$ as the second-longest path in the deterministic temporal network. Like the previous path, this path has a task-wise worst-case duration of $\mathbb{I}_{P_2}^\top \delta = 12.11$ and a path-wise worst-case duration of $\phi(x^3; P_2) = 10.61$. This path therefore also satisfies condition (iii) of Step 3(b), and we set $s = 3$.

In $Step\ 3(b)$, we identify $P_3 = \{1, 3, 5, 6\}$ as the third-longest path in the deterministic temporal network. This path has a task-wise worst-case duration of $\mathbb{I}_{P_3}^\top \delta = 9.86$. This path therefore satisfies condition (i) of Step 3(b), and we terminate with the optimal allocation $x^* = (0, 0.50, 0, 0.36, 0.14, 0)^\top$ and its worst-case makespan $\tau^* = 10.61$.

6.5 Upper Bounds

Consider a task path $P \in \overline{\mathcal{P}}$, together with a partition $\{P_r\}_{r=1}^R$ that satisfies $\bigcup_{r=1}^R P_r = P$ and $P_r \cap P_q = \emptyset$ for all $r \neq q$. According to (A2), we can bound P's worst-case duration $\phi(x; P)$ from above by $\sum_{r=1}^R \phi(x; P_r)$. Intuitively, this is the case because $\sum_{r=1}^R \phi(x; P_r)$ predicts different worst-case realizations of ξ for each block P_r, whereas $\phi(x; P)$ considers the same worst-case realization for all tasks in P. If we partition all paths $P \in \overline{\mathcal{P}}$ in this way, we obtain an upper bound on the optimal value of $\mathcal{ARTN}$. The granularity of the path partitions trades off the quality of the bound with the size of the associated bounding problem. If we use singleton partitions $\{\{i\}\}_{i \in P}$ for each path $P \in \overline{\mathcal{P}}$, for example, the associated optimization problem can be solved efficiently as a deterministic resource allocation problem with task durations $\phi(x; \{i\})$, $i \in V$. However, this approximation is very crude since it allows each task to attain its worst-case duration individually. At the other extreme, we recover $\mathcal{ARTN}$ if we employ single-block partitions $\{P\}$ for each path $P \in \overline{\mathcal{P}}$. In the following, we present an algorithm that iteratively advances

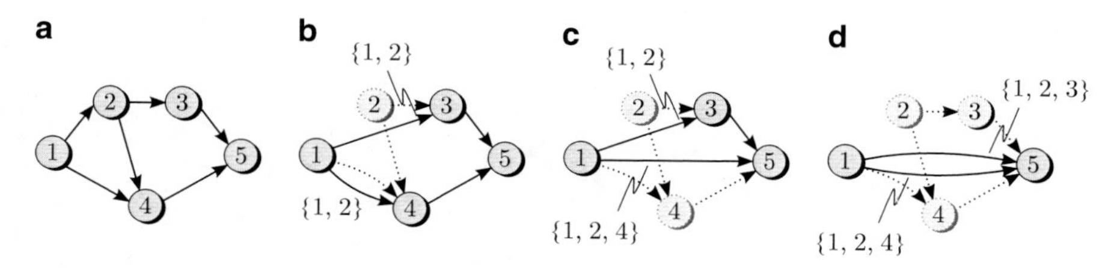

Fig. 6.5 Bounding graphs for the temporal network in (**a**). *Dotted nodes* (arcs) represent redundant variables (constraints) in the bounding problem

from singleton partitions to single-block partitions. We illustrate this idea with an example.

Example 6.5.1. Consider the temporal network in Fig. 6.5a. Assume that $\phi(x;\{5\}) = 0$, that is, task 5 has duration zero, and fix a resource allocation $x \in X$. Due to (A1) and (A2), the objective value of $\mathcal{ARTN}$ is the maximum of $\phi(x;\{1,2,3\})$ and $\phi(x;\{1,2,4\})$. We can bound this value from above if we replace the worst-case duration $\phi(x;P)$ of both paths $P \in \{\{1,2,3\},\{1,2,4\}\}$ with $\sum_{i \in P} \phi(x;\{i\})$. To calculate this bound, let $y \in \mathbb{R}^5_+$ denote the vector of task start times. We minimize y_5 subject to

$$y_2 \geq y_1 + \phi(x;\{1\}), \qquad y_3 \geq y_2 + \phi(x;\{2\}), \qquad y_4 \geq y_1 + \phi(x;\{1\}),$$
$$y_4 \geq y_2 + \phi(x;\{2\}), \qquad y_5 \geq y_3 + \phi(x;\{3\}), \qquad y_5 \geq y_4 + \phi(x;\{4\}).$$

This problem contains one constraint for each precedence in Fig. 6.5a. By construction, y_5 exceeds $\phi(x;\{1\}) + \phi(x;\{2\}) + \phi(x;\{3\})$ and $\phi(x;\{1\}) + \phi(x;\{2\}) + \phi(x;\{4\})$. We thus conclude that y_5 bounds $\mathcal{ARTN}$ from above.

The upper bound relies on the assumption that different tasks can attain different worst-case durations. To obtain a tighter bound, we coarsen the path partitions. We can achieve this by replacing the precedence $(1,2)$ in Fig. 6.5a with the two new precedences shown in Fig. 6.5b. The labels attached to these precedences list the tasks that need to be processed between the corresponding components of y. To calculate the new upper bound, we minimize y_5 subject to

$$y_3 \geq y_1 + \phi(x;\{1,2\}), \qquad y_4 \geq y_1 + \phi(x;\{1,2\}),$$
$$y_5 \geq y_3 + \phi(x;\{3\}), \qquad y_5 \geq y_4 + \phi(x;\{4\})$$

and the constraints corresponding to the dotted arcs in Fig. 6.5b. In the figure, dotted arcs lie on paths that are not inclusion-maximal, and property (A1) allows us to ignore the associated precedences. By construction, y_5 exceeds $\phi(x;\{1,2\}) + \phi(x;\{3\})$ and $\phi(x;\{1,2\}) + \phi(x;\{4\})$. Hence, y_5 still bounds $\mathcal{ARTN}$ from above. Our new bound is at least as tight as the old one since $\phi(x;\{1,2\}) \leq \phi(x;\{1\}) + \phi(x;\{2\})$. Note that the components of y cannot be interpreted as task start times anymore.

We now replace the labeled arc $(1,4)$ in Fig. 6.5b with the new labeled arc in Fig. 6.5c. To obtain the new upper bound, we minimize y_5 subject to

$$y_3 \geq y_1 + \phi(x;\{1,2\}), \quad y_5 \geq y_1 + \phi(x;\{1,2,4\}), \quad y_5 \geq y_3 + \phi(x;\{3\}).$$

Since y_5 exceeds $\phi(x;\{1,2\}) + \phi(x;\{3\})$ and $\phi(x;\{1,2,4\})$, it bounds $\mathcal{ARTN}$ from above. Again, the new upper bound is at least as tight as the previous one since $\phi(x;\{1,2,4\}) \leq \phi(x;\{1,2\}) + \phi(x;\{4\})$.

If we replace the labeled arc $(1,3)$ in Fig. 6.5c, then we obtain the graph in Fig. 6.5d. The associated bounding problem minimizes y_5 subject to

$$y_5 \geq y_1 + \phi(x;\{1,2,3\}), \qquad y_5 \geq y_1 + \phi(x;\{1,2,4\}).$$

This problem is equivalent to $\mathcal{ARTN}$. Note that for the path $\{1,2,3,5\}$, we iteratively generated the partitions $\{\{1\},\{2\},\{3\}\}$ in Fig. 6.5a, $\{\{1,2\},\{3\}\}$ in Fig. 6.5b and c and $\{\{1,2,3\}\}$ in Fig. 5.6d.

We now formalize the approach. To simplify the exposition, we assume that $\phi(x;\{n\}) = 0$ for all $x \in X$, that is, the sink node of the network has duration zero. This can always be achieved by introducing a dummy task.

For a temporal network $G = (V, E)$, we define a sequence of *bounding graphs* $G_1, G_2, \ldots$ as follows. Each bounding graph $G_t = (V, E_t)$ is directed and acyclic with nodes V and labeled arcs E_t. The arcs are of the form (j, k, P_{jk}), where $j, k \in V$ and the label P_{jk} satisfies $P_{jk} \subseteq V \setminus \{n\}$. There can be multiple arcs between j and k as long as they have different labels. The networks in Fig. 6.5 constitute bounding graphs if we attach the label $\{j\}$ to the each unlabeled arc from j to k.

We associate with G_t the following *bounding problem*:

$$\begin{aligned} \underset{x,y}{\text{minimize}} \quad & y_n \\ \text{subject to} \quad & x \in X, \;\; y \in \mathbb{R}^n_+ \\ & y_k - y_j \geq \phi(x; P_{jk}) \qquad \forall (j,k,P_{jk}) \in E_t. \end{aligned} \tag{$\mathcal{UARTN}_t$}$$

$\mathcal{UARTN}_t$ assigns a variable y_j to every node $j \in V$. The constraints ensure that y_k exceeds y_j by at least $\phi(x; P_{jk})$ time units if $(j, k, P_{jk}) \in E_t$. In Example 6.5.1 we formulated $\mathcal{UARTN}_t$ for the four bounding graphs in Fig. 6.5.

For a bounding graph G_t, we say that $P \in \overline{\mathcal{P}}$ is an *induced path* if every feasible solution (x, y) to G_t's bounding problem satisfies $y_n \geq \phi(x; P)$. To obtain an upper bound on $\mathcal{ARTN}$, we are interested in bounding graphs that induce all paths $P \in \overline{\mathcal{P}}$. Formally, we define the set of induced paths as

$$\mathcal{P}(G_t) = \Big\{ P \in \overline{\mathcal{P}} : \exists \{(i_r, i_{r+1}, P_r)\}_{r=1}^{R} \subseteq E_t \\ \text{such that } i_{R+1} = n \text{ and } (P \setminus \{n\}) = \bigcup_{r=1}^{R} P_r \Big\}.$$

Hence, $P \in \mathcal{P}(G_t)$ if the tasks in $P \setminus \{n\}$ are contained in the union of arc labels on a path in G_t that ends at the sink node n. Intuitively, y_n exceeds $\phi(x; P)$ because there is a partition $\{P_r\}_{r=1}^{R}$ of $P \setminus \{n\}$ such that $y_n \geq \sum_{r=1}^{R} \phi(x; P_r)$. Note that we can ignore the sink node n in this consideration since its duration is zero. The following lemma makes this argument explicit.

Lemma 6.5.1 (Induced Paths) *If $P \in \mathcal{P}(G_t)$, then any feasible solution (x, y) to $\mathcal{UARTN}_t$ satisfies $y_n \geq \phi(x; P)$.*

Proof. By definition of $\mathcal{P}(G_t)$, there is $\{(i_r, i_{r+1}, P_r)\}_{r=1}^{R} \subseteq E_t$ with $i_{R+1} = n$ and $(P \setminus \{n\}) = \bigcup_{r=1}^{R} P_r$. We thus have

$$y_n \overset{\text{(a)}}{\geq} y_n - y_{i_1} = \sum_{r=1}^{R} (y_{i_{r+1}} - y_{i_r}) \overset{\text{(b)}}{\geq} \sum_{r=1}^{R} \phi(x; P_r) \overset{\text{(c)}}{\geq} \phi(x; P \setminus \{n\}) \overset{\text{(d)}}{=} \phi(x; P),$$

where (a) follows from nonnegativity of y, (b) from the fact that (x, y) is feasible in $\mathcal{UARTN}_t$, and (c) and (d) from (A1), (A2) and $\phi(x; \{n\}) = 0$. □

As an illustration of induced paths, consider the path $\{1, 2, 4, 5\}$ in Example 6.5.1. It is induced by G_1 via $\{(1, 2, \{1\}), (2, 4, \{2\}), (4, 5, \{4\})\}$, by G_2 via $\{(1, 4, \{1, 2\}), (4, 5, \{4\})\}$, and by G_3 and G_4 via $\{(1, 5, \{1, 2, 4\})\}$, see Fig. 6.5. Lemma 6.5.1 implies that the objective value of any feasible solution (x, y) to $\mathcal{UARTN}_t$ provides an upper bound on the worst-case makespan of x with respect to all induced task paths. We conclude that $\mathcal{UARTN}_t$ bounds $\mathcal{ARTN}$ from above if $\overline{\mathcal{P}} \subseteq \mathcal{P}(G_t)$.

An initial upper bound on $\mathcal{ARTN}$ is obtained from $\mathcal{UARTN}_1$ where

$$G_1 = (V, E_1) \quad \text{with} \quad E_1 = \{(j, k, \{j\}) : (j, k) \in E\}. \tag{6.8}$$

$\mathcal{UARTN}_1$ comprises one constraint for every arc $(j, k, P_{jk}) \in E_1$. Since E_1 contains $|E|$ arcs, $\mathcal{UARTN}_1$ is a tractable optimization problem. The following lemma shows that $\mathcal{UARTN}_1$ bounds $\mathcal{ARTN}$ from above.

Lemma 6.5.2 (Initial Bound) *$\overline{\mathcal{P}} \subseteq \mathcal{P}(G_1)$ for G_1 defined in (6.8).*

Proof. Consider any path $P = \{i_1 = 1, i_2, \ldots, i_{R+1} = n\} \in \overline{\mathcal{P}}$ with $(i_r, i_{r+1}) \in E$ for $r = 1, \ldots, R$. For $P_r = \{i_r\}$, $r = 1, \ldots, R$, we have $\{(i_r, i_{r+1}, P_r)\}_{r=1}^{R} \subseteq E_1$ and $(P \setminus \{n\}) = \bigcup_{r=1}^{R} P_r$, so that $P \in \mathcal{P}(G_1)$. □

Figure 6.5a visualizes G_1 for the temporal network in Example 6.5.1. The initial bounding graph approximates the worst-case duration $\phi(x; P)$ of every path $P \in \overline{\mathcal{P}}$ by the duration $\sum_{i \in P} \phi(x; \{i\})$ of the singleton partition $\{\{i\}\}_{i \in P}$. If this approximation is tight, then $\mathcal{UARTN}_1$ and $\mathcal{ARTN}$ are equivalent. This is the case, for example, if all task durations depend on disjoint parts of ξ that are not related to each other through Ξ. In general, however, $\phi(x; P) < \sum_{i \in P} \phi(x; \{i\})$, and the optimal value of $\mathcal{UARTN}_1$ constitutes a strict upper bound on the optimal value of $\mathcal{ARTN}$.

By suitably transforming the graph G_1, we can coarsen the path partitions to tighten the upper bound provided by $\mathcal{UARTN}_1$.

Definition 6.5.1 (Replacements) For a bounding graph $G_t = (V, E_t)$ we construct $G_{t+1} = (V, E_{t+1})$ via the following two types of replacements.

1. *Predecessor replacement:* G_{t+1} *results from a* predecessor replacement *of* $(j, k, P_{jk}) \in E_t$ *if* $j \neq 1$ *and*

$$E_{t+1} = E_t \setminus \{(j, k, P_{jk})\} \cup \bigcup_{\substack{i \in V, P_{ij} \in \mathcal{P}: \\ (i, j, P_{ij}) \in E_t}} \{(i, k, P_{ij} \cup P_{jk})\}.$$

2. *Successor replacement:* G_{t+1} *results from a* successor replacement *of* $(j, k, P_{jk}) \in E_t$ *if* $k \neq n$ *and*

$$E_{t+1} = E_t \setminus \{(j, k, P_{jk})\} \cup \bigcup_{\substack{l \in V, P_{kl} \in \mathcal{P}: \\ (k, l, P_{kl}) \in E_t}} \{(j, l, P_{jk} \cup P_{kl})\}.$$

The two replacements are illustrated in Figs. 6.6 and 6.7. We call an arc $(j, k, P_{jk}) \in E_t$ *replaceable* if it qualifies for either of the two replacements. The application of a replacement to $(j, k, P_{jk}) \in E_t$ reduces the approximation error for every path $P \in \mathcal{P}(G_t)$ whose partition $\{P_r\}_{r=1}^{R}$ contains the block P_{jk}. At the same time, however, the number of arcs in the resulting bounding graph (and hence the size of the bounding problem) typically increases. In Example 6.5.1

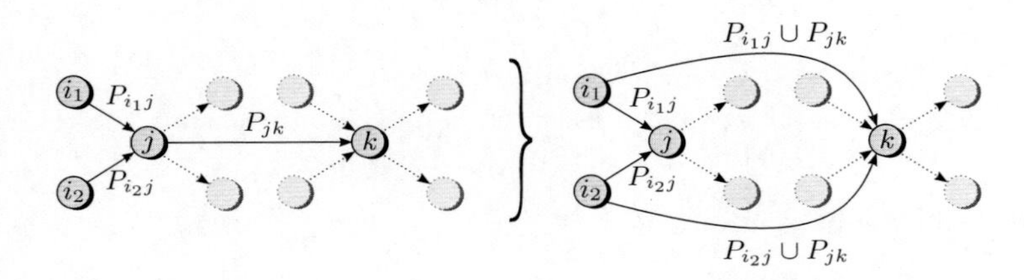

Fig. 6.6 Predecessor replacement of (j, k, P_{jk}) with two predecessor nodes

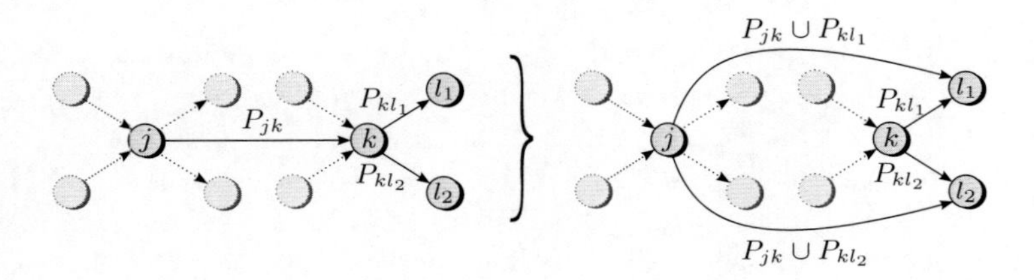

Fig. 6.7 Successor replacement of (j, k, P_{jk}) with two successor nodes

we applied successor replacements to $(1, 2, \{1\}) \in E_1$, $(1, 4, \{1, 2\}) \in E_2$ and $(1, 3, \{1, 2\}) \in E_3$. As the result of a replacement, some nodes and/or arcs in the bounding graph may become redundant, see Fig. 6.5. We will identify such redundancies at the end of this section.

From now on, we assume that $\{G_t\}_t$ is a sequence of bounding graphs where G_1 is defined in (6.8) and $G_2, G_3, \ldots$ result from an iterated application of replacements in the sense of Definition 6.5.1. In this case, the label P_{jk} of an arc $(j, k, P_{jk}) \in E_t$ contains precisely the tasks on a path from j to k (excluding task k) in the temporal network G.

Lemma 6.5.3 *For each arc $(j, k, P_{jk}) \in E_t$ the temporal network G contains a directed path $\{(l_r, l_{r+1})\}_{r=1}^{R} \subseteq E$ with $R \geq 1$, $(l_1, l_{R+1}) = (j, k)$ and $P_{jk} = \{l_1, \ldots, l_R\}$.*

Proof. We prove the assertion by induction on t. By construction of G_1, the assertion holds for $t = 1$. Assume now that the assertion holds for G_t and that G_{t+1} results from a predecessor replacement of $(j, k, P_{jk}) \in E_t$ (an analogous argument can be made for successor replacements). According to Definition 6.5.1, any new arc in $E_{t+1} \setminus E_t$ must be of the form (i, k, P_{ik}), and E_t must contain an arc $(i, j, P_{ij}) \in E_t$ with $P_{ij} \cup P_{jk} = P_{ik}$. Since the assertion holds for G_t, G contains directed paths $\{(l_r, l_{r+1})\}_{r=1}^{R}, \{(l'_r, l'_{r+1})\}_{r=1}^{R'} \subseteq E$ with $(l_1, l_{R+1}) = (i, j)$, $(l'_1, l'_{R'+1}) = (j, k)$, $P_{ij} = \{l_1, \ldots, l_R\}$ and $P_{jk} = \{l'_1, \ldots, l'_{R'}\}$. Since $l_{R+1} = l'_1$, we can connect both paths to prove the assertion for (i, k, P_{ik}). Since the arc $(j, k, P_{jk}) \in E_t$ was chosen arbitrarily, the assertion of the lemma follows. □

The next lemma shows that replacements preserve the upper bound property.

Lemma 6.5.4 (Bound Preservation) *If $\overline{\mathcal{P}} \subseteq \mathcal{P}(G_t)$, then $\overline{\mathcal{P}} \subseteq \mathcal{P}(G_{t+1})$.*

Proof. Choose any path $P \in \overline{\mathcal{P}}$. By assumption, $P \in \mathcal{P}(G_t)$, that is, there exists a set of arcs $\{(i_r, i_{r+1}, P_r)\}_{r=1}^{R} \subseteq E_t$ with $i_{R+1} = n$ and $(P \setminus \{n\}) = \bigcup_{r=1}^{R} P_r$. We show that $P \in \mathcal{P}(G_{t+1})$. Assume that G_{t+1} results from a predecessor replacement of $(j, k, P_{jk}) \in E_t$; the proof is widely parallel for successor replacements.

If $(j, k) \neq (i_r, i_{r+1})$ for all $r \in \{1, \ldots, R\}$, then $P \in \mathcal{P}(G_{t+1})$ is vacuously satisfied. Hence, assume that $(j, k) = (i_s, i_{s+1})$ for some $s \in \{1, \ldots, R\}$. Since 1 is the unique source of G (see Section 1.1) and $P \in \overline{\mathcal{P}}$, we have $1 \in P$. Lemma 6.5.3 then implies that $i_1 = 1$. Hence, $s \neq 1$ since (i_1, i_2, P_1) does not qualify for a predecessor replacement. Let $i'_r = i_r$ for $r = 1, \ldots, s-1$ and $i'_r = i_{r+1}$ for $r = s, \ldots, R$. Similarly, let $P'_r = P_r$ for $r = 1, \ldots, s-2$ (if $s > 2$), $P'_{s-1} = P_{s-1} \cup P_s$ and $P'_r = P_{r+1}$ for $r = s, \ldots, R-1$. We have that $\left\{(i'_r, i'_{r+1}, P'_r)\right\}_{r=1}^{R-1} \subseteq E_{t+1}$, $i'_R = n$ and $(P \setminus \{n\}) = \bigcup_{r=1}^{R-1} P'_r$, which ensures that $P \in \mathcal{P}(G_{t+1})$. Since P was chosen arbitrarily, the assertion follows. □

We can now prove that the proposed replacements result in a monotonically non-increasing, convergent sequence of upper bounds on $\mathcal{ARTN}$.

Proposition 6.5.1 *Let (x^t, y^t) denote an optimal solution to $\mathcal{UARTN}_t$. Then:*

(a) For every t, x^t is a feasible allocation in $\mathcal{ARTN}$ and y^t_n is an upper bound on the worst-case makespan of x^t in $\mathcal{ARTN}$.
(b) There is $T \in \mathbb{N}$ such that there are no replaceable arcs in G_T. For this T, x^T is an optimal allocation in $\mathcal{ARTN}$ and y^T_n is the worst-case makespan of x^T in $\mathcal{ARTN}$.
(c) The sequence $\{y^t_n\}_{t=1}^T$ is monotonically non-increasing.

Proof. By construction, x^t constitutes a feasible allocation for every t. Due to Lemma 6.5.1, assertion (a) is therefore satisfied if $\overline{\mathcal{P}} \subseteq \mathcal{P}(G_t)$ for every t. Employing Lemmas 6.5.2 and 6.5.4, this follows by induction on t.

As for (b), we recall that G_1 is acyclic. Hence, we can relabel the nodes of G_1 such that all $(j, k, P_{jk}) \in E_1$ satisfy $j < k$. Every replacement removes one arc $(j, k, P_{jk}) \in E_t$, $t = 1, 2, \ldots$, and adds less than $|E_t|$ arcs (i, l, P_{il}) with $i \leq j$ and $l \geq k$, where one of these inequalities is strict. Since all $(j, k, P_{jk}) \in E_t$ satisfy $1 \leq j, k \leq n$, there is $T \in \mathbb{N}$ such that there are no replaceable arcs in G_T.

All arcs in E_T are of the form $(1, n, P_{1n})$ for some $P_{1n} \subseteq V \setminus \{n\}$ since otherwise, further replacements would be possible. Hence, $\mathcal{UARTN}_T$ is equivalent to

$$\min_{x \in X} \max_{(1,n,P_{1n}) \in E_T} \phi(x; P_{1n}).$$

We have $\overline{\mathcal{P}} \subseteq \{P_{1n} \in \mathcal{P} : (1, n, P_{1n}) \in E_T\} \subseteq \mathcal{P}$ due to Lemma 6.5.3 and part (a) of this proof. Hence, $\mathcal{UARTN}_T$ is equivalent to $\mathcal{ARTN}$, and claim (b) follows.

To prove (c), we first show that if (x, y) is feasible in $\mathcal{UARTN}_t$, $t \in \{1, \ldots, T-1\}$, then it is also feasible in $\mathcal{UARTN}_{t+1}$. Assume that G_{t+1} is obtained from a predecessor replacement of $(j, k, P_{jk}) \in E_t$. The argument is widely parallel for successor replacements. $\mathcal{UARTN}_{t+1}$ results from $\mathcal{UARTN}_t$ by replacing the constraint $y_k - y_j \geq \phi(x; P_{jk})$ with new constraints of the form $y_k - y_i \geq \phi(x; P_{ij} \cup P_{jk})$ for $i \in V$ and $P_{ij} \subseteq V \setminus \{n\}$ with $(i, j, P_{ij}) \in E_t$. These new constraints are less restrictive, however, because

$$y_k - y_i = (y_k - y_j) + (y_j - y_i) \overset{\text{(i)}}{\geq} \phi(x; P_{ij}) + \phi(x; P_{jk}) \overset{\text{(ii)}}{\geq} \phi(x; P_{ij} \cup P_{jk}).$$

Here, (i) follows from the fact that (x, y) is feasible in $\mathcal{UARTN}_t$, while (ii) is due to (A2). Hence, (x, y) is feasible in $\mathcal{UARTN}_{t+1}$, too. Since $\mathcal{UARTN}_t$ and $\mathcal{UARTN}_{t+1}$ share the same objective function, assertion (c) follows. □

Proposition 6.5.1 provides the justification for Algorithm 6.

Algorithm 6 does not prescribe the choice of any specific replacement. We will discuss a selection scheme below. Before that, we summarize the following algorithm properties which are a direct consequence of Proposition 6.5.1.

Corollary 6.5.1 *Algorithm 6 terminates with an optimal resource allocation x^* in $\mathcal{ARTN}$ and its worst-case makespan y^*_n. Moreover, $\{x^t\}_{t=1}^T$ represents a sequence*

Algorithm 6 Convergent upper bounds on $\mathcal{ARTN}$.

1. **Initialization.** Construct G_1 as defined in (6.8). Set $t = 1$.
2. **Bounding Problem.** Find an optimal solution (x^t, y^t) to $\mathcal{UARTN}_t$.
3. **Replacement.** Choose a replaceable arc $(j, k, P_{jk}) \in E_t$.
   (a) *If there is no such arc*, terminate: $x^* = x^t$ is an optimal allocation in $\mathcal{ARTN}$ and $y_n^* = y_n^t$ is the worst-case makespan of x^* in $\mathcal{ARTN}$.
   (b) *Otherwise*, construct G_{t+1} by applying a replacement to arc (j, k, P_{jk}), set $t \to t + 1$ and go to Step 2.

of feasible allocations in $\mathcal{ARTN}$ and $\{y_n^t\}_{t=1}^T$ a monotonically non-increasing sequence of upper bounds on their objective values in $\mathcal{ARTN}$.

By combining Algorithms 4 and 6, we obtain monotonically convergent lower and upper bounds on the optimal value of $\mathcal{ARTN}$, together with feasible allocations $x^t \in X$ whose worst-case makespans are bracketed by these bounds. This provides us with feasible allocations that converge to the optimal allocation and whose suboptimality can be quantified at any iteration.

The tractability assumption (A3) allows us to reduce the set of meaningful replacement candidates in Step 3 of Algorithm 6 as follows.

Proposition 6.5.2 *Assume that (A3) holds, and let (x^t, y^t) denote any optimal solution to $\mathcal{UARTN}_t$. We have:*

(a) If $y_k^t - y_j^t > \phi(x^t; P_{jk})$ for some *replaceable arc $(j, k, P_{jk}) \in E_t$, then $\mathcal{UARTN}_{t+1}$ with G_{t+1} obtained from G_t by replacing (j, k, P_{jk}) has an optimal value of y_n^t, too.*
(b) If $y_k^t - y_j^t > \phi(x^t; P_{jk})$ for all *replaceable arcs $(j, k, P_{jk}) \in E_t$, then $\mathcal{UARTN}_s$ with $s > t$ and G_s obtained from G_t by any sequence of replacements has an optimal value of y_n^t, too.*

Remark 6.5.1. According to assertion (a), replacing any arc $(j, k, P_{jk}) \in E_t$ that satisfies the described condition leads to the same upper bound as $\mathcal{UARTN}_t$. Since we intend to reduce this bound, we may disregard all such replacement candidates in Step 3 of Algorithm 6. Part (b) describes a condition under which x^t is the optimal allocation and y_n^t the optimal value of $\mathcal{ARTN}$.

Proof of Proposition 6.5.2 Assume that (a) is false, that is, $y_k^t - y_j^t > \phi(x^t; P_{jk})$, but there is a feasible solution (x^{t+1}, y^{t+1}) to $\mathcal{UARTN}_{t+1}$ that has an objective value smaller than y_n^t. From the argumentation in the proof of Proposition 6.5.1 (c) we know that (x^t, y^t) is feasible in $\mathcal{UARTN}_{t+1}$. Due to (A3),

$$(x^\lambda, y^\lambda) = \lambda(x^{t+1}, y^{t+1}) + (1 - \lambda)(x^t, y^t) \quad \text{for } \lambda \in (0, 1]$$

is also feasible for $\mathcal{UARTN}_{t+1}$ and has an objective value smaller than y_n^t. We show that for small λ, (x^λ, y^λ) is feasible in $\mathcal{UARTN}_t$, too. Since

$E_t \setminus E_{t+1} = \{(j, k, P_{jk})\}$, we only need to show that $y^\lambda_k - y^\lambda_j \geq \phi(x^\lambda; P_{jk})$. For sufficiently small λ, this follows from continuity of $\phi(\cdot; P_{jk})$ in its first component, which is a consequence of (A3), and the fact that $y^t_k - y^t_j > \phi(x^t; P_{jk})$. Since $\mathcal{UARTN}_t$ and $\mathcal{UARTN}_{t+1}$ share the same objective function, this implies that (x^t, y^t) is not optimal for $\mathcal{UARTN}_t$. Thus, our assumption is false and (a) must be true.

As for (b), let us now assume that $y^t_k - y^t_j > \phi(x^t; P_{jk})$ for all replaceable arcs $(j, k, P_{jk}) \in E_t$. In this case, assertion (a) guarantees that (x^t, y^t) remains optimal for G_{t+1} if G_{t+1} results from applying one replacement to G_t. Assume that G_{t+1} results from a predecessor replacement of $(j, k, P_{jk}) \in E_t$ (the proof for successor replacements is analogous). We then have

$$(y^t_k - y^t_i) = (y^t_k - y^t_j) + (y^t_j - y^t_i) \overset{\text{(i)}}{>} \phi(x^t; P_{ij}) + \phi(x^t; P_{jk})$$
$$\overset{\text{(ii)}}{\geq} \phi(x^t; P_{ij} \cup P_{jk}) \qquad \forall (i, j, P_{ij}) \in E_t,$$

where (i) follows from the assumption and (ii) is due to (A2). Hence, the condition described in assertion (b) is satisfied for all new arcs $(i, k, P_{ij} \cup P_{jk}) \in E_{t+1}$ as well. An iterated application of this argument shows that assertion (b) remains valid for $\mathcal{UARTN}_s$ with G_s obtained from applying any sequence of predecessor and/or successor replacements to G_t. This implies that $\mathcal{UARTN}_s$ has an optimal value of y^t_n, and thus the claim follows. □

$\mathcal{UARTN}_t$ may have several optimal solutions, and the conditions in Proposition 6.5.2 may only be satisfied for some of them. If an optimal solution (x^t, y^t) to $\mathcal{UARTN}_t$ does not satisfy the condition in Proposition 6.5.2 (a) for $(j, k, P_{jk}) \in E_t$, then we can use y^t_n to check whether other optimal solutions (x', y') satisfy the condition. Indeed, this is the case if

$$\max_{\substack{x \in X, \\ y \in \mathbb{R}^n_+}} \{(y_k - y_j) - \phi(x; P_{jk}) \,:\, y_n = y^t_n,\ y_q - y_p \geq \phi(x; P_{pq})$$
$$\forall (p, q, P_{pq}) \in E_t\} > 0. \tag{6.9}$$

Similarly, Proposition 6.5.2 (b) implies that x^t is an optimal allocation for $\mathcal{ARTN}$ if all replacement candidates $(j, k, P_{jk}) \in E_t$ satisfy the inequality (6.9). Unfortunately, evaluating the left-hand side of the inequality (6.9) is as difficult as solving $\mathcal{UARTN}_t$, and it is prohibitive to compute it for all $(j, k, P_{jk}) \in E_t$. If we fix x to x^t and optimize the left-hand side of the inequality (6.9) only over y, however, the maximization can be computed in time $\mathcal{O}(|E_t|)$ by a combined forward and backward calculation, see [DH02]. In this case, however, we might not identify all replacement candidates that satisfy the conditions of Proposition 6.5.2.

Although Proposition 6.5.2 reduces the set of potential replacement candidates, it provides no criterion for selecting specific arcs to be replaced. Ideally, one would choose a replacement that leads to the largest reduction of the upper bound. This approach is computationally prohibitive, however, since it requires the solution of bounding problems for all replacement candidates. Likewise, "first fit" approaches are unsuited due to similar reasons as in Sect. 6.4. Instead, one may choose a replacement for G_t that leads to the largest reduction of the upper bound when x is fixed to the optimal allocation of $\mathcal{UARTN}_t$. Like the optimization of the left-hand side of the inequality (6.9) for fixed x, this evaluation requires time $\mathcal{O}(|E_t|)$ and can hence be implemented efficiently. At the same time, however, this selection scheme is likely to lead to better results than naive "first fit" approaches.

Let us now consider the issue of redundant nodes and arcs in the bounding graphs G_t. We call an arc $(j, k, P_{jk}) \in E_t$ *redundant* if it can be removed from E_t without changing the set of induced paths $\mathcal{P}(G_t)$. The following proposition lists sufficient conditions for redundancy.

Proposition 6.5.3 *An arc* $(j, k, P_{jk}) \in E_t$ *is redundant if one of the following conditions is met:*

1. *There is another arc* (j, k, P'_{jk}) *with* $P_{jk} \subseteq P'_{jk}$, $P_{jk} \neq P'_{jk}$.
2. *Node* j *has no incoming arcs in* G_t *and* $j \neq 1$.
3. *Node* k *has no outgoing arcs in* G_t *and* $k \neq n$.

The proof of this proposition is straightforward, and we omit it for the sake of brevity. Proposition 6.5.3 allows us to identify redundant nodes as well: node $i \in V$ is redundant in G_t if all of its incoming and outgoing arcs are redundant.

We close this section with an example that illustrates Algorithm 6.

Example 6.5.2. Consider again the problem instance from Examples 6.3.2 and 6.4.1. We generate upper bounds on the optimal objective value of this problem with Algorithm 6.

We start with *Step 1*, where we construct the bounding graph G_1 shown in Fig. 6.8, upper left. Note that we added a dummy sink node 7 and an artificial precedence between nodes 6 and 7 so that the last task (i.e., task 7) has duration zero. Since none of the arcs in the bounding graph G_1 satisfies the conditions of

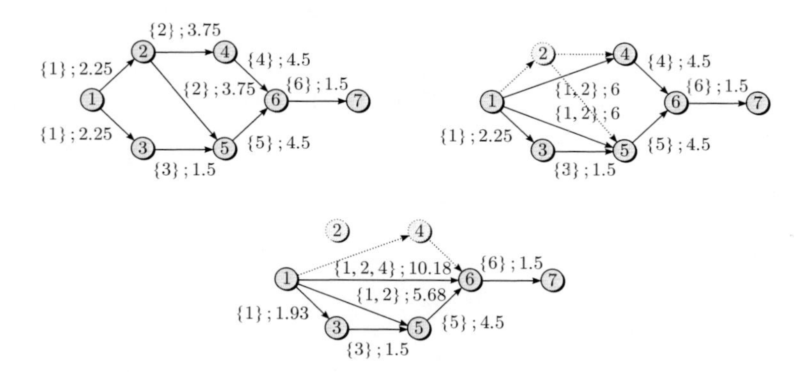

Fig. 6.8 Bounding graphs generated by Algorithm 6. The *upper left*, *upper right* and *lower charts* visualize the bounding graphs in iteration $t = 1$, $t = 2$ and $t = 3$, respectively. Attached to each arc $(i, j, P_{ij}) \in E_t$ is its label P_{ij} and its worst-case duration $\phi(x^t; P_{ij})$

Proposition 6.5.3, we cannot identify any arc or node in G_1 as redundant. We set $t = 1$.

In *Step 2* we solve the upper bound problem $\mathcal{UARTN}_1$:

$$
\begin{aligned}
\underset{x,y,\lambda,\gamma}{\text{minimize}} \quad & y_7 \\
\text{subject to} \quad & x \in \mathbb{R}^6_+, \quad y \in \mathbb{R}^7_+, \quad \lambda \in \mathbb{R}^8_+, \quad \gamma \in \mathbb{R}^8_+ \\
& y_2 \geq y_1 + 2(1 - x_1) + \lambda_1^1/2 + \gamma^1, \quad \lambda_1^1 + \gamma^1 \geq 2(1 - x_1), \\
& y_3 \geq y_1 + 2(1 - x_1) + \lambda_1^2/2 + \gamma^2, \quad \lambda_1^2 + \gamma^2 \geq 2(1 - x_1), \\
& y_4 \geq y_2 + 5(1 - x_2) + \lambda_2^3/2 + \gamma^3, \quad \lambda_2^3 + \gamma^3 \geq 5(1 - x_2), \\
& y_5 \geq y_2 + 5(1 - x_2) + \lambda_2^4/2 + \gamma^4, \quad \lambda_2^4 + \gamma^4 \geq 5(1 - x_2), \\
& y_5 \geq y_3 + 1(1 - x_3) + \lambda_3^5/2 + \gamma^5, \quad \lambda_3^5 + \gamma^5 \geq 1(1 - x_3), \\
& y_6 \geq y_4 + 4(1 - x_4) + \lambda_4^6/2 + \gamma^6, \quad \lambda_4^6 + \gamma^6 \geq 4(1 - x_4), \\
& y_6 \geq y_5 + 3(1 - x_5) + \lambda_5^7/2 + \gamma^7, \quad \lambda_5^7 + \gamma^7 \geq 3(1 - x_5), \\
& y_7 \geq y_6 + 1(1 - x_6) + \lambda_6^8/2 + \gamma^8, \quad \lambda_6^8 + \gamma^8 \geq 1(1 - x_6), \\
& x_i \leq 1/2 \;\; \forall i \in \{1, \ldots, 6\}, \quad \sum_{i=1}^{6} x_i \leq 1.
\end{aligned}
$$

The optimal allocation is given by $x^1 = (0.25, 0.5, 0, 0.25, 0, 0)^\top$, and the optimal task start schedule is $y^1 = (0, 2.25, 4.5, 6, 6, 10.5, 12)^\top$. The estimated worst-case makespan is 12.

According to Proposition 6.5.2, we should not replace the arc $(1, 3, \{1\}) \in E_1$ because $\phi(x^1; \{1\}) = 2.25$ but $y_3^1 - y_1^1 = 4.5$. If we apply the extended check described in (6.9), we see that we should not replace the arc $(3, 5, \{3\}) \in E_1$ either. For ease of exposition, we use a "first-fit" approach here and apply a forward replacement to the arc $(1, 2, \{1\}) \in E_1$ in *Step 3*. The new bounding graph G_2 is visualized in Fig. 6.8, upper right. Note that the arcs $(2, 4, \{2\})$ and $(2, 5, \{2\})$ in E_2 satisfy the second condition of Proposition 6.5.3 and are therefore redundant. As a result, node 2 is redundant as well. We set $t = 2$.

Back in *Step 2*, we solve the upper bound problem $\mathcal{UARTN}_2$:

$$
\begin{aligned}
\underset{x,y,\lambda,\gamma}{\text{minimize}} \quad & y_7 \\
\text{subject to} \quad & x \in \mathbb{R}^6_+, \quad y \in \mathbb{R}^6_+, \quad \lambda \in \mathbb{R}^9_+, \quad \gamma \in \mathbb{R}^7_+ \\
& y_4 \geq y_1 + 2(1 - x_1) + \lambda_1^1/2 + 5(1 - x_2) + \lambda_2^1/2 + \gamma^1, \\
& \lambda_1^1 + \gamma^1 \geq 2(1 - x_1), \quad \lambda_2^1 + \gamma^1 \geq 5(1 - x_2), \\
& y_5 \geq y_1 + 2(1 - x_1) + \lambda_1^2/2 + 5(1 - x_2) + \lambda_2^2/2 + \gamma^2,
\end{aligned}
$$

$$\lambda_1^2 + \gamma^2 \geq 2(1 - x_1), \quad \lambda_2^2 + \gamma^2 \geq 5(1 - x_2),$$
$$y_3 \geq y_1 + 2(1 - x_1) + \lambda_1^3/2 + \gamma^3, \quad \lambda_1^3 + \gamma^3 \geq 2(1 - x_1),$$
$$y_5 \geq y_3 + 1(1 - x_3) + \lambda_3^4/2 + \gamma^4, \quad \lambda_3^4 + \gamma^4 \geq 1(1 - x_3),$$
$$y_6 \geq y_4 + 4(1 - x_4) + \lambda_4^5/2 + \gamma^5, \quad \lambda_4^5 + \gamma^5 \geq 4(1 - x_4),$$
$$y_6 \geq y_5 + 3(1 - x_5) + \lambda_5^6/2 + \gamma^6, \quad \lambda_5^6 + \gamma^6 \geq 3(1 - x_5),$$
$$y_7 \geq y_6 + 1(1 - x_6) + \lambda_6^7/2 + \gamma^7, \quad \lambda_6^7 + \gamma^7 \geq 1(1 - x_6),$$
$$x_i \leq 1/2 \;\; \forall i \in \{1, \ldots, 6\}, \quad \sum_{i=1}^{6} x_i \leq 1.$$

The optimal allocation is $x^2 = (0.25, 0.5, 0, 0.25, 0, 0)^\top$, and the optimal vector y^2 is given by $y_1^2 = 0$, $y_3^2 = 4.5$, $y_4^2 = y_5^2 = 6$, $y_6^2 = 10.5$ and $y_7^2 = 12$. The estimated worst-case makespan is 12. Note that neither the optimal allocation nor the upper bound changed. This is due to the fact that our uncertainty set allows two tasks to attain their worst-case durations simultaneously.

As before, Proposition 6.5.2 indicates that we should not replace the arcs $(1, 3, \{1\})$ and $(3, 5, \{3\})$ in E_2. Instead, we apply a forward replacement to the arc $(1, 4, \{1, 2\}) \in E_2$ in *Step 3*. The new bounding graph G_3 is visualized in Fig. 6.8, lower chart. Due to the replacement, the arc $(4, 6, \{4\}) \in E_3$ and node 4 have become redundant. We set $t = 3$.

Back in *Step 2*, we solve the upper bound problem $\mathcal{UARTN}_3$:

$$\begin{aligned}
\underset{x,y,\lambda,\gamma}{\text{minimize}} \quad & y_7 \\
\text{subject to} \quad & x \in \mathbb{R}_+^6, \;\; y \in \mathbb{R}_+^5, \;\; \lambda \in \mathbb{R}_+^9, \;\; \gamma \in \mathbb{R}_+^6 \\
& y_6 \geq y_1 + 2(1 - x_1) + \lambda_1^1/2 + 5(1 - x_2) + \lambda_2^1/2 + 4(1 - x_4) + \lambda_4^1 + \gamma^1, \\
& \lambda_1^1 + \gamma^1 \geq 2(1 - x_1), \quad \lambda_2^1 + \gamma^1 \geq 5(1 - x_2), \; \lambda_4^1 + \gamma^1 \geq 4(1 - x_4), \\
& y_5 \geq y_1 + 2(1 - x_1) + \lambda_1^2/2 + 5(1 - x_2) + \lambda_2^2/2 + \gamma^2, \\
& \lambda_1^2 + \gamma^2 \geq 2(1 - x_1), \quad \lambda_2^2 + \gamma^2 \geq 5(1 - x_2), \\
& y_3 \geq y_1 + 2(1 - x_1) + \lambda_1^3/2 + \gamma^3, \quad \lambda_1^3 + \gamma^3 \geq 2(1 - x_1), \\
& y_5 \geq y_3 + 1(1 - x_3) + \lambda_3^4/2 + \gamma^4, \quad \lambda_3^4 + \gamma^4 \geq 1(1 - x_3), \\
& y_6 \geq y_5 + 3(1 - x_5) + \lambda_5^5/2 + \gamma^5, \quad \lambda_5^5 + \gamma^5 \geq 3(1 - x_5), \\
& y_7 \geq y_6 + 1(1 - x_6) + \lambda_6^6/2 + \gamma^6, \quad \lambda_6^6 + \gamma^6 \geq 1(1 - x_6), \\
& x_i \leq 1/2 \;\; \forall i \in \{1, \ldots, 6\}, \quad \sum_{i=1}^{6} x_i \leq 1.
\end{aligned}$$

The optimal allocation is $x^3 \approx (0.36, 0.5, 0, 0.14, 0, 0)^\top$, and the optimal vector y^3 is given by $y_1^3 = 0$, $y_3^3 \approx 1.94$, $y_5^3 \approx 5.68$, $y_6^3 \approx 10.18$ and $y_7^3 \approx 11.68$. The estimated worst-case makespan is approximately 11.68.

Proposition 6.5.2 indicates that we should not replace the arcs $(1, 3, \{1\})$ and $(3, 5, \{3\})$ in E_3. We can proceed by applying a forward replacement to the arc $(1, 6, \{1, 2, 4\})$. For the sake of brevity, we omit the remaining steps of our bounding approach.

6.6 Numerical Results for Random Test Instances

We investigate the performance of the bounding technique and compare it with the decision rule approximations reviewed in Sect. 6.2.2. To this end, we use the RANGEN algorithm described in [DVH03] to generate 100 random instances of problem $\mathcal{RTN}$ of size $n \in \{100, 200, 300\}$ and order strength 0.25, 0.5 and 0.75. The *order strength* of a network $G = (V, E)$ denotes the fraction of all $n(n-1)/2$ theoretically possible precedences between the nodes in V that are enforced through the arcs in E (either directly or via transitivity), see [DH02]. Table 6.1 summarizes the median numbers of inclusion-maximal paths for each instance class. Note that this number increases with the instance size and the order strength. We expect instances with a larger number of paths to be more challenging to solve with the bounding approach.

We solve the resource allocation problem outlined in Example 6.3.2, that is, we assume a single resource and task durations

$$d_i(x; \xi) = d_i^0 (1 - x_i)(1 + \xi_i) \qquad \text{for } i \in V,$$

where d_i^0 denotes the nominal task duration, x_i the amount of the resource that is assigned to task i, and ξ_i the uncertainty inherent to the task duration. We sample d_i^0 uniformly from the interval $[1, 10]$ and set

$$X = \left\{x \in \mathbb{R}_+^n \,:\, x \leq (1/2)\,\mathrm{e},\ \mathrm{e}^\top x \leq \beta\right\}$$

$$\text{and} \quad \Xi = \left\{\xi \in \mathbb{R}_+^n \,:\, \xi \leq (1/2)\,\mathrm{e},\ \mathrm{e}^\top \xi \leq \gamma\right\}.$$

Table 6.1 Numbers of inclusion-maximal paths for the generated instance classes

n	0.25	0.50	0.75
100	1,158	14,940	1,929,456
200	7,275	390,715	3,134,873,127
300	22,893	3,477,994	608,740,179,463

Each class is described by its network size (row) and its order strength (column)

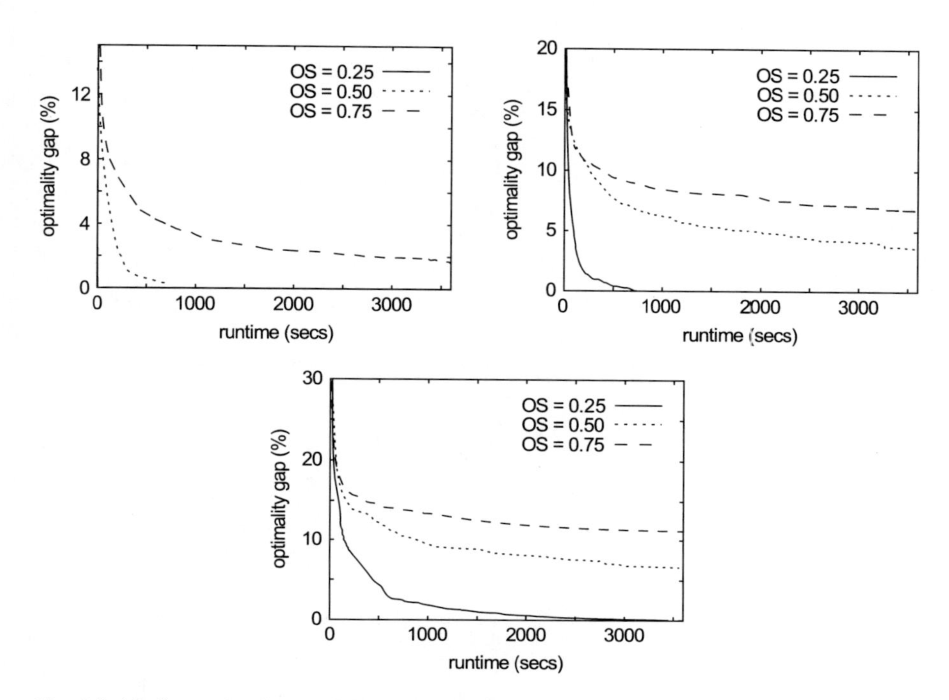

Fig. 6.9 Median optimality gaps of the bounding approach as functions of the runtime. The graphs show the results for instances of size 100 (*upper left*), 200 (*upper right*) and 300 (*lower graph*). In the first graph, the optimality gap for OS = 0.25 vanishes so quickly that the curve cannot be seen

Thus, the duration of task i can fall below or exceed its nominal duration d_i^0 by 50%, depending on the resource allocation and the realization of the uncertain parameter vector ξ. We choose the resource budget β such that 10% of all tasks can be sped up to their minimal durations. Likewise, we select the uncertainty budget γ such that on average 10% of the tasks on each inclusion-maximal path can attain their worst-case durations.

Even though they constitute linear programs, the resulting instances of $\mathcal{ARTN}$ are difficult to solve with a standard optimizer. Indeed, for instances with 100 tasks and an order strength of 0.5, $\mathcal{ARTN}$ already contains more than 345,000 variables and 235,000 constraints on average. To bound the optimal value of $\mathcal{ARTN}$, we run the algorithms from Sects. 6.4 and 6.5 in parallel for 1 h. We solve all intermediate optimization problems with IBM ILOG CPLEX 12.1 on a 2.53 GHz Intel Core 2 Duo computer. Figure 6.9 visualizes the resulting optimality gaps as functions of the computation time. As expected, instances with a large number of tasks and a high order strength are more difficult to solve. Apart from instances with 300 tasks and an order strength of 0.75, however, the optimality gaps after 1 h are all below 10%. Moreover, more than 90% of the instances of three classes (100 tasks with an order strength of 0.25; 100 tasks with an order strength of 0.5; 200 tasks with an order strength of 0.25) are solved within the time limit.

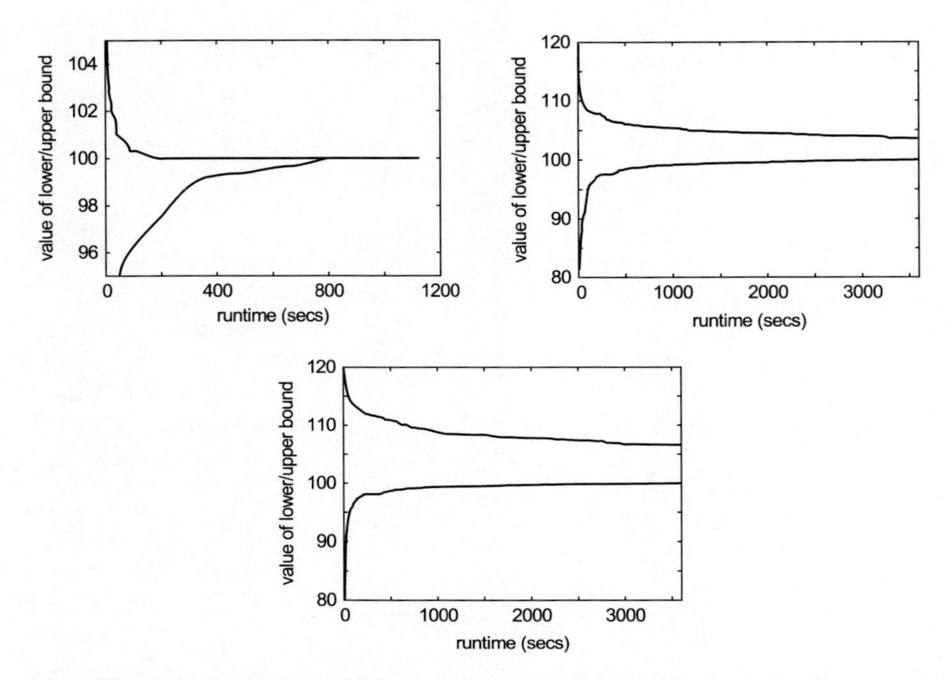

Fig. 6.10 Median lower and upper bounds of the bounding approach as functions of the runtime. The graphs visualize the results for instances of size 100 (*upper left*), 200 (*upper right*) and 300 (*lower graph*) and an order strength of 0.5. The objective values are normalized so that the lower bound after 1 h evaluates to 100

We now investigate the individual contributions of the upper and lower bounds to the optimality gaps in Fig. 6.9. To this end, Fig. 6.10 presents the upper and lower bounds as functions of the runtime for instances with an order strength of 0.5. For instances with 200 and 300 tasks, the lower bound improves rapidly in the beginning but fails to prove optimality within the time limit. Indeed, the graphs reveal that the upper and lower bounds improve throughout the computation, although the progress slows down after some time.

We now compare the results of the bounding approach with the decision rule approximations outlined in Sect. 6.2.2. We were unable to solve the affine decision rule approximations for any of the test instances within the time limit of 1 h. Indeed, the optimization models for instances with 100 tasks and an order strength of 0.25 already contain more than 140,000 variables and 130,000 constraints on average. We therefore restrict each affine decision rule $y_j(\xi)$, $j \in V$, to depend on a small number of random variables ξ_i associated with the task durations d_i of predecessor tasks i of j. The results are presented in Table 6.2. Note that experiments in which less than 50% of the instances could be solved within 1 h do not allow the calculation of median values; the corresponding entries in the table are therefore labeled "n/a". As expected, affine decision rules perform better than constant decision rules, and the approximation quality of the affine decision rules improves with the number

Table 6.2 Computational results for constant decision rules (CDR) and affine decision rules over 5 (ADR-5) and 10 (ADR-10) random variables

n	CDR			ADR-5			ADR-10		
	0.25	0.50	0.75	0.25	0.50	0.75	0.25	0.50	0.75
100	32.16%	36.91%	36.77%	19.57%	26.47%	28.00%	15.85%	22.11%	23.17%
	0.62	0.98	1.28	1.92	8.14	11.32	6.98	39.11	42.82
200	30.10%	31.09%	33.30%	22.14%	22.35%	25.47%	18.74%	19.16%	22.10%
	4.36	7.16	9.47	18.63	270.45	443.74	91.8	1,100.53	1,562.63
300	26.65%	27.40%	30.95%	19.64%	21.65%	22.60%	17.56%	n/a	n/a
	12.53	23.69	37.75	181.71	2,062.24	2,612.82	717.77	n/a	n/a

The results are grouped as in Table 6.1. The first value of each entry shows the median percentage by which the decision rule approximation exceeds the final upper bound of the bounding approach, while the second value presents the median runtime in seconds

Table 6.3 Median optimality gaps (first row) and runtimes (second row) of the bounding approach for different resource budgets β and uncertainty budgets γ

Nominal	Budget β		Budget γ	
	20%	30%	20%	30%
5.85%	0.54%	0.00%	8.04%	10.21%
3,600.0	3,600.0	74.22	3,600.0	3,600.0

The nominal test set uses the values for β and γ that are described in the beginning of the section. The remaining test sets increase one of the budgets by a factor of 2 (20%) or 3 (30%)

of considered random variables. However, the results are consistently dominated by the bounding approach presented in this chapter. The results in Table 6.2 could be improved by using piecewise affine decision rules, but in this case the allowed computation time would have to be increased considerably.

We close with an analysis of the impact of the resource budget β and the uncertainty budget γ on the bounding approach. To this end, Table 6.3 shows the optimality gaps and runtimes for different values of β and γ over instances of size 200 and an order strength of 0.5. We see that the bounding scheme works best if β is large and γ is small. An empirical inspection revealed that in this case, the intermediate resource allocations change less between consecutive iterations of the lower and upper bounds. We suspect that this "allocation stability" allows the bounding algorithm to progress faster.

6.7 Case Study: VLSI Design

We now apply the bounding technique to a circuit sizing problem with process variations. For a survey of optimization problems in circuit design, see [BKPH05].

An important problem in circuit design is to select the gate sizes in a circuit with the goal to optimally balance three conflicting objectives: operating speed, circuit size and power consumption. Loosely speaking, larger gate sizes increase the circuit size and power consumption, but they reduce the gate delays. We can model a circuit as a temporal network with gates as tasks and interconnections between gates as precedences. The duration of task $i \in V$ refers to the delay of gate i. The makespan of the network corresponds to the delay of the overall circuit, which in turn is inversely proportional to the circuit's operating speed. A resource allocation assigns sizes to all gates in the circuit.

The maximization of circuit speed, subject to constraints on power consumption and circuit size, can be cast as a deterministic resource allocation problem that is defined on a temporal network. In practice, however, a circuit represents only one component of a larger system, and its eventual operating speed depends on adjacent circuits (that are outside the model). Hence, one commonly imposes a lower bound on the circuit speed and minimizes the circuit size instead. For the sake of simplicity, we ignore power consumption here. The deterministic problem then becomes

$$\begin{aligned} \underset{x,y}{\text{minimize}} \quad & \sum_{i \in V} A_i x_i \\ \text{subject to} \quad & x \in \mathbb{R}^n_+, \; y \in \mathbb{R}^n_+ \\ & y_j \geq y_i + d_i(x) \qquad \forall (i,j) \in E, \\ & y_n + d_n(x) \leq T, \\ & x \in [\underline{x}, \overline{x}]. \end{aligned} \tag{6.10}$$

Here, x_i represents the size of gate i (with positive lower and upper bounds $\underline{x}_i$ and $\overline{x}_i$, respectively) and $A_i x_i$ the area occupied by gate i. Assuming that the circuit has a unique sink n (see Sect. 1.1), $y_n + d_n(x)$ denotes the delay of the overall circuit. We require that this quantity must not exceed some target value T. Note that for some values of T, the problem may be infeasible.

In the following, we employ a resistor–capacitor model for the gate delays:

$$d_i(x) = 0.69 \frac{R_i}{x_i} \left(C_i^{\text{int}} x_i + \sum_{j:(i,j) \in E} C_j^{\text{in}} x_j \right) \quad \text{for } i \in V, \; x \in X, \tag{6.11}$$

where R_i, C_i^{int} and C_i^{in} denote the driving resistance, intrinsic capacitance and input capacitance of gate i, respectively [BKPH05].

Variations in the manufacturing process entail that the factual gate sizes deviate from the selected target sizes x by some random, zero-mean noise $\xi \in \mathbb{R}^n$. If this noise is small compared to x, then we can express the resulting gate delays $d_i(x+\xi)$, $i \in V$, by a first-order Taylor approximation:

$$d_i(x;\xi) = d_i(x) + \left[\nabla d_i(x)\right]^\top \xi \quad \text{for } i \in V.$$

Process variations exhibit nonnegative correlations [SNLS05]. We can account for such correlations by using an ellipsoidal uncertainty set:

$$\Xi = \left\{ \xi \in \mathbb{R}^n \,:\, \exists u \in \mathbb{R}^l \,.\, \xi = \Sigma u, \; \|u\|_2 \leq 1 \right\} \quad \text{with } \Sigma \in \mathbb{R}^{n \times l}_+. \tag{6.12}$$

We thus seek to optimize the following variant of $\mathcal{RTN}$:

$$\inf_{x\in[\underline{x},\overline{x}]} \sup_{\xi\in\Xi} \inf_{y\in Y(x,\xi)} \left\{\sum_{i\in V} A_i x_i \;:\; y_n + d_n(x;\xi) \leq T\right\}. \tag{6.13}$$

For a suitable ϕ (see Sect. 6.3), this results in the following variant of $\mathcal{ARTN}$:

$$\begin{array}{lll} \underset{x}{\text{minimize}} & \sum_{i\in V} A_i x_i & \\ \text{subject to} & x \in \mathbb{R}^n_+ & \\ & \phi(x;P) \leq T & \forall P \in \mathcal{P}, \\ & x \in [\underline{x},\overline{x}]\,. & \end{array} \tag{6.14}$$

Again, problem (6.14) may be infeasible if T is chosen too small. An inspection of Sects. 6.4 and 6.5 reveals that we can apply the bounding approach to problem (6.14) if we allow the bounds to attain values on the extended real line $\mathbb{R} \cup \{\infty\}$. A lower bound of ∞ signalizes that problem (6.14) is infeasible, while an upper bound of ∞ indicates that the determined gate sizes x may violate the target value T for the overall circuit delay. The following result provides us with a conservative reformulation of the problem (6.13):

Proposition 6.7.1 *For d and Ξ defined in (6.11) and (6.12), let*

$$\phi(x;P) = \mathbb{I}_P^\top d(x) + \left\|\Sigma^\top\Big(\sum_{i\in P}[\nabla d_i(x)]^+\Big)\right\|_2 + \left\|\Sigma^\top\Big(\sum_{i\in P}[\nabla d_i(x)]^-\Big)\right\|_2, \tag{6.15}$$

where

$$[f(x)]^+ = \sum_{i:\alpha_i>0} \alpha_i \prod_j (x_j)^{\beta_{ij}} \quad \textit{for} \quad f(x) = \sum_i \alpha_i \prod_j (x_j)^{\beta_{ij}}$$

and $[f(x)]^-$ defined analogously for i with $\alpha_i < 0$. If X has a tractable representation, then the problem (6.14), with ϕ defined in (6.15), satisfies (A1)–(A3) and constitutes a conservative reformulation of the problem (6.13).

Proof. It follows from [SNLS05] that ϕ as defined in (6.15) satisfies condition (6.5) on p. 115 and (A3). It remains to be shown that ϕ satisfies (A1) and (A2). For $x \in X$ and $P \subseteq V$, we introduce the following notation:

$$\varphi^+(x,P) = \left\|\Sigma^\top\Big(\sum_{i\in P}[\nabla d_i(x)]^+\Big)\right\|_2 \text{ and } \varphi^-(x,P) = \left\|\Sigma^\top\Big(\sum_{i\in P}[\nabla d_i(x)]^-\Big)\right\|_2.$$

As for (A1), we need to show that

$$\begin{aligned} \phi(x;P') &= \mathbb{I}_{P'}^\top d(x) + \varphi^+(x,P') + \varphi^-(x;P') \\ &\geq \mathbb{I}_P^\top d(x) + \varphi^+(x;P) + \varphi^-(x;P) \;=\; \phi(x;P) \end{aligned}$$

Table 6.4 ISCAS 85 benchmark circuits

Circuit	# tasks	# precedences	# task paths
C432	196	336	83,926
C499	243	408	9,440
C880	443	729	8,642
C1355	587	1,064	4,173,216
C1908	913	1,498	729,056
C2670	1,426	2,076	679,954
C3540	1,719	2,939	28,265,874
C5315	2,485	4,386	1,341,305
C6288	2,448	4,800	1,101,055,638
C7552	3,719	6,144	726,494

for all $x \in X$ and $P \subset P' \subseteq V$. Note that $\mathbb{I}_{P'}^{\top} d(x) \geq \mathbb{I}_{P}^{\top} d(x)$ since $\mathbb{I}_{P'} \geq \mathbb{I}_{P}$ and $d(x) \geq 0$ for all $x \in X$. We show that $\varphi^{+}(x; P') \geq \varphi^{+}(x; P)$ and $\varphi^{-}(x; P') \geq \varphi^{-}(x; P)$. The first inequality follows from the fact that Σ is element-wise nonnegative and $\left[\nabla d_i(x)\right]^{+} \geq 0$ for all $i \in V$. The second inequality follows from the positive homogeneity of norms and the fact that $\left[\nabla d_i(x)\right]^{-} \leq 0$ for all $i \in V$.

Property (A2) is satisfied if

$$\begin{aligned}\phi(x; P) + \phi(x; P' \setminus P) &= \mathbb{I}_{P}^{\top} d(x) + \varphi^{+}(x; P) + \varphi^{-}(x; P) \\ &\quad + \mathbb{I}_{[P' \setminus P]}^{\top} d(x) + \varphi^{+}(x; P' \setminus P) + \varphi^{-}(x; P' \setminus P) \\ &\geq \mathbb{I}_{P'}^{\top} d(x) + \varphi^{+}(x, P') + \varphi^{-}(x; P') = \phi(x; P')\end{aligned}$$

for all $x \in X$ and $P \subset P' \subseteq V$. Note that $\mathbb{I}_{P}^{\top} d(x) + \mathbb{I}_{[P' \setminus P]}^{\top} d(x) = \mathbb{I}_{P'}^{\top} d(x)$. Also, we have

$$\varphi^{+}(x; P) + \varphi^{-}(x; P) + \varphi^{+}(x; P' \setminus P) + \varphi^{-}(x; P' \setminus P) \geq \varphi^{+}(x, P') + \varphi^{-}(x; P')$$

by the triangle inequality. □

We use Proposition 6.7.1 to determine robust gate sizes for the ISCAS 85 benchmark circuits.[2] To this end, we set $(\underline{x}_i, \overline{x}_i) = (1, 16)$ and select the circuit parameters A_i, R_i, C_i^{int} and C_i^{in} according to the Logical Effort model [BKPH05, SSH99]. We set the target delay T to 130% of the minimal circuit delay in absence of process variations. For ease of exposition, we assume independent process variations, that is, Σ is a diagonal matrix. We set the diagonal elements of Σ to 25% of the gate sizes determined by the deterministic model (6.10).

The data in Table 6.4 specifies the temporal networks corresponding to the ISCAS 85 benchmark circuits. For a circuit with $|V|$ tasks and $|\overline{\mathcal{P}}|$

[2]ISCAS 85 benchmark circuits: `http://www.cbl.ncsu.edu/benchmarks`.

Table 6.5 Results for the circuits from Table 6.4

Circuit	First it.	After 25 its.	After 50 its.	Reduction
C432	34.13%	*Solved after 11 its.*		*24.48%*
	0:03	1:03		
C499	148.82%	12.31%	8.96%	*42.89%*
	0:12	27:35	128:30	
C880	16.78%	2.31%	0.70%	*11.16%*
	0:11	2:44	8:39	
C1355	113.16%	*Solved after 24 its.*		*52.95%*
	0:17	17:31		
C1908	37.05%	11.37%	6.90%	*18.13%*
	1:17	6:58	21:06	
C2670	14.62%	1.61%	1.02%	*11.09%*
	0:51	24:03	99:35	
C3540	37.66%	9.19%	7.40%	*20.50%*
	4:22	16:31	56:06	
C5315	15.23%	4.30%	2.29%	*10.33%*
	6:56	30:39	52:37	
C6288	68.24%	3.40%	2.52%	*39.07%*
	6:33	45:09	69:08	
C7552	11.03%	*Solved after 12 its.*		*5.01%*
	5:54	15:08		

Columns 2–4 present the optimality gaps and computation times (mins:secs) after 1, 25 and 50 iterations of the bounding approach, respectively. The last column quantifies the reduction in overall circuit size if we use the bounding approach instead of constant decision rules (see Sect. 6.2.2)

inclusion-maximal task paths, the path-wise model (6.14) can be reformulated as a geometric program with $1 + |V| + 2\left|\overline{\mathcal{P}}\right|$ variables and $3\left|\overline{\mathcal{P}}\right|$ constraints, see [BKPH05,SNLS05]. Due to the choice of ϕ in (6.15), the Jacobian of the constraints is dense. In view of the cardinality of $\overline{\mathcal{P}}$ in the benchmark circuits (see Table 6.4), a direct solution of problem (6.14) is prohibitive.

We now use the bounding approach to solve problem (6.14) for the benchmark circuits. We terminate the algorithm after 50 iterations of the lower and upper bound procedures. Since the lower bound requires the investigation of a potentially large number of task paths (see Step 3(b) of Algorithm 5), we limit its computation time per iteration to the time required by the upper bound. All results are generated with CONOPT 3 on an Intel Xeon architecture with 2.83 GHz.[3] We employ warm starts for the calculation of both lower and upper bounds, which significantly reduces the computational effort.

Table 6.5 presents the optimality gaps after 1, 25 and 50 iterations. It also documents the reduction in overall circuit size when we use the bounding approach (for 50 iterations) instead of a model with constant decision rules (see Sect. 6.2.2).

[3]CONOPT homepage: `http://www.conopt.com`.

We remark that the choice of Ξ and ϕ in (6.12) and (6.15) implies that constant and affine decision rules result in the same solutions. Although the initial optimality gaps can be large, the bounding approach reduces them to reasonable values after a few iterations. Moreover, the computational effort remains modest for all considered problem instances. Finally, we see that the bounding approach can lead to significant reductions in overall circuit size.

6.8 Conclusion

This chapter studied robust resource allocations in temporal networks. We considered a problem formulation which assumes that the task durations are uncertain and that resource allocations are evaluated in view of their worst-case makespan. We showed that the resulting optimization problem is $\mathcal{NP}$-hard. We computed convergent bounds on its optimal objective value, as well as feasible resource allocations whose objective values are bracketed by these bounds.

It would be interesting to extend the solution procedure to renewable and doubly constrained resources. Indeed, Sect. 2.1 lists some application domains (e.g., scheduling of production processes and microprocessors) that impose additional restrictions on the consumption rate of resources. Such constraints result in non-convex problems that render the bounding approach computationally prohibitive. Instead, one could design a branch-and-bound algorithm that branches upon violations of the additional restrictions. For every node in the resulting branch-and-bound tree, the incurred worst-case makespan can be bounded with the method presented in this chapter.

References

[AK89] Adlakha VG, Kulkarni VG (1989) A classified bibliography of research on stochastic PERT networks: 1966–1987. INFOR 27(3):272–296

[AG03] Alizadeh F, Goldfarb D (2003) Second-order cone programming. Math Program 95(1):3–51

[AP05] Ardagna D, Pernici B (2005) Global and local QoS constraints guarantee in web service selection. In: Proceedings of the IEEE international conference on web services (ICWS), IEEE Computer Society, Orlando, Florida, pp 805–806

[ADEH99] Artzner P, Delbaen F, Eber J-M, Heath D (1999) Coherent measures of risk. Math Finance 9(3):203–228

[AZ07] Atamtürk A, Zhang M (2007) Two-stage robust network flow and design under demand uncertainty. Oper Res 55(4):662–673

[Ave01] Averbakh I (2001) On the complexity of a class of combinatorial optimization problems with uncertainty. Math Program 90(2):263–272

[BSS06] Bazaraa MS, Sherali HD, Shetty CM (2006) Nonlinear programming—theory and algorithms, 3rd edn. Wiley, New York

[Ben06] Benati S (2006) An optimization model for stochastic project networks with cash flows. Comput Manag Sci 3(4):271–284

[BTN00] Ben-Tal A, Nemirovski A (2000) Robust solutions of linear programming problems contaminated with uncertain data. Math Program 88(3):411–424

[BTGGN04] Ben-Tal A, Goryashko A, Guslitzer E, Nemirovski A (2004) Adjustable robust solutions of uncertain linear programs. Math Program 99(2):351–376

[BTGN09] Ben-Tal A, El Ghaoui L, Nemirovski A (2009) Robust optimization. Princeton University Press, Princeton

[Ber73] Berk KN (1973) A central limit theorem for m-dependent random variables with unbounded m. Ann Probab 1(2):352–354

[Ber07] Bertsekas DP (2007) Dynamic programming and optimal control, vol 2. Athena Scientific, Nashua

[BT96] Bertsekas DP, Tsitsiklis J (1996) Neuro-dynamic programming. Athena Scientific, Nashua

[BP05] Bertsimas D, Popescu I (2005) Optimal inequalities in probability theory: a convex optimization approach. SIAM J Optim 15(3):780–804

[BS04] Bertsimas D, Sim M (2004) The price of robustness. Oper Res 52(1):35–53

[BS06] Bertsimas D, Sim M (2006) Tractable approximations to robust conic optimization problems. Math Program 107(1–2):5–36

[BS03] Bertsimas D, Sim M (2007) Robust discrete optimization and network flows. Math Program 98(1–3):49–71

W. Wiesemann, *Optimization of Temporal Networks under Uncertainty*,
Advances in Computational Management Science 10, DOI 10.1007/978-3-642-23427-9,

[BNT02] Bertsimas D, Natarajan K, Teo C-P (2002) Applications of semidefinite optimization in stochastic project scheduling. Technical Report, Singapore—MIT Alliance

[BM95] Birge JR, Maddox MJ (1995) Bounds on expected project tardiness. Oper Res 43(5):838–850

[BEP+96] Błażewicz J, Ecker KH, Pesch E, Schmidt G, Węglarz J (1996) Scheduling computer and manufacturing processes, 2nd edn. Springer, Heidelberg

[BKPH05] Boyd SP, Kim S-J, Patil DD, Horowitz MA (2005) Digital circuit optimization via geometric programming. Oper Res 53(6):899–932

[Bru07] Brucker P (2007) Scheduling algorithms, 5th edn. Springer, Heidelberg

[BDM+99] Brucker P, Drexl A, Möhring R, Neumann K, Pesch E (1999) Resource-constrained project scheduling: notation, classification, models, and methods. Eur J Oper Res 112(1):3–41

[Bus95] Buss AH (1995) Sequential experimentation for estimating the optimal delay of activities in PERT networks. In: Proceedings of the 1995 winter simulation conference, Arlington, Virginia, pp 336–340

[BR97] Buss AH, Rosenblatt MJ (1997) Activity delay in stochastic project networks. Oper Res 45(1):126–139

[CC05] Calafiore G, Campi MC (2005) Uncertain convex programs: randomized solutions and confidence levels. Math Program 102(1):25–46

[CC06] Calafiore G, Campi MC (2006) The scenario approach to robust control design. IEEE Trans Automat Contr 51(5):742–753

[CCGP07a] Cardellini V, Casalicchio E, Grassi V, Lo Presti F (2007) Flow-based service selection for web service composition supporting multiple QoS classes. In: Proceedings of the IEEE international conference on web services (ICWS 2007), IEEE Computer Society, Salt Lake City, Utah, pp 743–750

[CCGP07b] Cardellini V, Casalicchio E, Grassi V, Lo Presti F (2007) Scalable service selection for web service composition supporting differentiated QoS classes. Research Report RR-07.59, Università di Roma Tor Vergata

[CSS07] Chen X, Sim M, Sun P (2007) A robust optimization perspective on stochastic programming. Oper Res 55(6):1058–1071

[CSSZ08] Chen X, Sim M, Sun P, Zhang J (2008) A linear-decision based approximation approach to stochastic programming. Oper Res 56(2):344–357

[CSST10] Chen W, Sim M, Sun J, Teo C-P (2010) From CVaR to uncertainty set: implications in joint chance constrained optimization. Oper Res 58(2):470–485

[CGS07] Cohen I, Golany B, Shtub A (2007) The stochastic time-cost tradeoff problem: a robust optimization approach. Networks 49(2):175–188

[DHV+95] Deckro RF, Hebert JE, Verdini WA, Grimsrud PH, Venkateshwar S (1995) Nonlinear time/cost tradeoff models in project management. Comput Ind Eng 28(2):219–229

[DY10] Delage E, Ye Y (2010) Distributionally robust optimization under moment uncertainty with application to data-driven problems. Oper Res 58(3):596–612

[DH02] Demeulemeester EL, Herroelen WS (2002) Project scheduling—a research handbook. Kluwer Academic, Dordrecht

[DDH93] Demeulemeester EL, Dodin B, Herroelen WS (1993) A random activity network generator. Oper Res 41(5):972–980

[DVH03] Demeulemeester EL, Vanhoucke M, Herroelen WS (2003) RanGen: a random network generator for activity-on-the-node networks. J Scheduling 6(1):17–38

[Den99] Deng S (1999) Stochastic models of energy commodity prices and their applications: mean-reversion with jumps and spikes. Technical Report, UC Berkeley

[Elm00] Elmaghraby SE (2000) On criticality and sensitivity in activity networks. Eur J Oper Res 127(2):220–238

[Elm05] Elmaghraby SE (2005) On the fallacy of averages in project risk management. Eur J Oper Res 165(2):307–313

[EH90] Elmaghraby SE, Herroelen WS (1990) The scheduling of activities to maximize the net present value of projects. Eur J Oper Res 49(1):35–49

[EK90] Elmaghraby SE, Kamburowski J (1990) On project representation and activity floats. Arab J Sci Eng 15(4B):626–637

[EK92] Elmaghraby SE, Kamburowski J (1992) The analysis of activity networks under generalized precedence relations (GPRs). Manag Sci 38(9):1245–1263

[Epp94] Eppstein D (1994) Finding the k shortest paths. In: IEEE symposium on foundations of computer science, Santa Fe, New Mexico, pp 154–165

[EI07] Erdoğan E, Iyengar G (2007) On two-stage convex chance constrained problems. Math Meth Oper Res 65(1):115–140

[EMS09] Erera AL, Morales JC, Savelsbergh M (2009) Robust optimization for empty repositioning problems. Oper Res 57(2):468–483

[FJMM07] Feige U, Jain K, Mahdian M, Mirrokni V (2007) Robust combinatorial optimization with exponential scenarios. In: Fischetti M, Williamson DP (eds) Integer programming and combinatorial optimization. Springer, Heidelberg, pp 439–453

[Fis70] Fishburn PC (1970) Utility theory for decision making. Wiley, New York

[FL04] Floudas CA, Lin X (2004) Continuous-time versus discrete-time approaches for scheduling of chemical processes: a review. Comput Chem Eng 28(11):2109–2129

[Ful61] Fulkerson DR (1961) A network flow computation for project cost curves. Manag Sci 7(2):167–178

[GJ79] Garey MR, Johnson DS (1979) Computers and intractability: a guide to the theory of NP-completeness. Freeman, New York

[GG06] Goel V, Grossmann IE (2006) A class of stochastic programs with decision dependent uncertainty. Math Program 108(2–3):355–394

[GS10] Goh J, Sim M (2010) Distributionally robust optimization and its tractable approximations. Oper Res 58(4):902–917

[GYTZ06] Gao A, Yang D, Tang S, Zhang M (2006) QoS-driven web service composition with inter service conflicts. In: Proceedings of the 8th Asia-Pacific conference on frontiers of WWW research and development (APWeb 2006), Harbin, China. Springer, Heidelberg, pp 121–132

[GH68] Gordon TJ, Hayward H (1968) Initial experiments with the cross-impact matrix method of forecasting. Futures 1(2):100–116

[Gri72] Grinold RC (1972) The payment scheduling problem. Nav Res Logist Quarterly 19(1):123–136

[Hag88] Hagstrom JN (1988) Computational complexity of PERT problems. Networks 18(2):139–147

[HR03] Heitsch H, Römisch W (2003) Scenario reduction algorithms in stochastic programming. Comput Optim Appl 24(2–3):187–206

[HS08] Henrion R, Strugarek C (2008) Convexity of chance constraints with independent random variables. Comput Optim Appl 41(2):263–276

[HKR09] Henrion R, Küchler C, Römisch W (2009) Scenario reduction in stochastic programming with respect to discrepancy distances. Comput Optim Appl 43(1):67–93

[HL04] Herroelen WS, Leus R (2004) Robust and reactive project scheduling: a review and classification of procedures. Int J Prod Res 42(8):1599–1620

[HL05] Herroelen WS, Leus R (2005) Project scheduling under uncertainty: survey and research potentials. Eur J Oper Res 165(2):289–306

[HDD97] Herroelen WS, Van Dommelen P, Demeulemeester EL (1997) Project network models with discounted cash flows – a guided tour through recent developments. Eur J Oper Res 100(1):97–121

[HK93] Hettich R, Kortanek KO (1993) Semi-infinite programming: theory, methods, and applications. SIAM Rev 35(3):380–429

[HPT00] Horst R, Pardalos PM, Thoai NV (2000) Introduction to global optimization, 2nd edn. Kluwer Academic, Dordrecht

[HCLC05] Huang Y, Chung J-Y, Li Y, Chao K-M (2005) A stochastic service composition model for business integration. In: Proceedings of the international conference on next generation web services practices (NWeSP 2005). IEEE Computer Society, Seoul, Korea, pp 418–428

[Iba80] Ibaraki T (1980) Approximate algorithms for the multiple-choice continuous knapsack problems. J Oper Res Soc Jpn 23(1):28–63

[JL05] Jaeger MC, Ladner H (2005) Improving the QoS of WS compositions based on redundant services. In: Proceedings of the international conference on next generation web services practices (NWeSP 2005), IEEE Computer Society, Seoul, Korea, pp 189–194

[JM99] Jain AS, Meeran S (1999) Deterministic job-shop scheduling: past, present and future. Eur J Oper Res 113(2):390–434

[JLF07] Janak SL, Lin X, Floudas CA (2007) A new robust optimization approach for scheduling under uncertainty: II. Uncertainty with known probability distribution. Comput Chem Eng 31(3):171–195

[JW00] Jørgensen T, Wallace SW (2000) Improving project cost estimation by taking into account managerial flexibility. Eur J Oper Res 127(2):239–251

[JWW98] Jonsbråten TW, Wets RJ-B, Woodruff DL (1998) A class of stochastic programs with decision dependent random elements. Ann Oper Res 82(1):83–106

[KW94] Kall P, Wallace SW (1994) Stochastic programming. Wiley, New York

[Kel61] Kelley JE (1961) Critical-path planning and scheduling: mathematical basis. Oper Res 9(3):296–320

[KC02] Kenyon C, Cheliotis G (2002) Architecture requirements for commercializing grid resources. In: Proceedings of the 11th IEEE international symposium on high performance distributed computing (HPDC 2002), IEEE Computer Society, Edinburgh, Scotland, pp 215–224

[KKMS08] Khandekar R, Kortsarz G, Mirrokni V, Salavatipour MR (2008) Two-stage robust network design with exponential scenarios. In: Halperin D, Mehlhorn K (eds) Algorithms—ESA 2008, Springer, Heidelberg, pp 589–600

[KBY+07] Kim S-J, Boyd SP, Yun S, Patil DD, Horowitz MA (2007) A heuristic for optimizing stochastic activity networks with applications to statistical digital circuit sizing. Optim Eng 8(4):397–430

[KD07] Kokash N, D'Andrea V (2007) Evaluating quality of web services: a risk-driven approach. In: Proceedings of the 10th international conference on business information systems (BIS 2007), Poznań, Poland, Springer, Heidelberg, pp 180–194

[KPK07] Korski J, Pfeuffer F, Klamroth K (2007) Biconvex sets and optimization with biconvex functions: a survey and extensions. Math Meth Oper Res 66(3):373–407

[KY97] Kouvelis P, Yu G (1997) Robust discrete optimization and its applications. Kluwer Academic, Dordrecht

[KWG] Kuhn D, Wiesemann W, Georghiou (2011) A Primal and dual linear decision rules in stochastic and robust optimization. Math Program 130(1):177–209

[KA86] Kulkarni VG, Adlakha VG (1986) Markov and Markov-regenerative PERT networks. Oper Res 34(5):769–781

[LKY05] Lakhal NB, Kobayashi T, Yokota H (2005) A failure-aware model for estimating and analyzing the efficiency of web services compositions. In: Proceedings of the 11th Pacific Rim international symposium on dependable computing (PRDC 2005), Hunan, China, pp 114–124

[LI08] Li Z, Ierapetritou M (2008) Process scheduling under uncertainty: review and challenges. Comput Chem Eng 32(4–5):715–727

[LLMS09] Liebchen C, Lübbecke ME, Möhring RH, Stiller S (2009) The concept of recoverable robustness, linear programming recovery, and railway applications. In: Ahuja RK, Möhring RH, Zaroliagis CD (eds) Robust and online large-scale optimization, Springer, Heidelberg, pp 1–27

[LP94] Limongelli C, Pirastu R (1994) Exact solution of linear equation systems over rational numbers by parallel p-adic arithmetic. RISC Report Series 94–25, University of Linz

[LJF04] Lin X, Janak SL, Floudas CA (2004) A new robust optimization approach for scheduling under uncertainty: I. Bounded uncertainty. Comput Chem Eng 28 (6–7):1069–1085

[LMS01] Ludwig A, Möhring RH, Stork F (2001) A computational study on bounding the makespan distribution in stochastic project networks. Ann Oper Res 102(1):49–64

[LA08] Luedtke J, Ahmed S (2008) A sample approximation approach for optimization with probabilistic constraints. SIAM J Optim 19(2):674–699

[MRCF59] Malcolm DG, Roseboom JH, Clark CE, Fazar W (1959) Application of a technique for research and development program evaluation. Oper Res 7(5):646–669

[Mar52] Markowitz HM (1952) Portfolio selection. J Finance 7(1):77–91

[MCWG95] Mas-Colell A, Whinston MD, Green JR (1995) Microeconomic theory. Oxford University Press, Oxford

[MN79] Meilijson I, Nádas A (1979) Convex majorization with an application to the length of critical paths. J Appl Probab 16(3):671–677

[Men02] Menascé DA (2002) QoS issues in web services. IEEE Internet Comput 6(6):72–75

[MM04] Milanovic N, Malek M (2004) Current solutions for web service composition. IEEE Internet Comput 8(6):51–59

[Mö01] Möhring RH (2001) Scheduling under uncertainty: bounding the makespan distribution. In: Alt H (ed) Conputational discrete mathematics, Springer, Heidelberg, pp 79–97

[MS00] Möhring R, Stork F (2000) Linear preselective policies for stochastic project scheduling. Math Meth Oper Res 52(3):501–515

[Neu79] Neumann K (1979) Recent advances in temporal analysis of GERT networks. Z Oper Res 23:153–177

[Neu99] Neumann K (1999) Scheduling of projects with stochastic evolution structure. In: Węglarz J (ed) Project scheduling: recent models, algorithms, and applications, Kluwer Academic, Dordrecht, pp 309–332

[NS06a] Nemirovski A, Shapiro A (2006) Convex approximations of chance constrained programs. SIAM J Optim 17(4):969–996

[NS06b] Nemirovski A, Shapiro A (2006) Scenario approximations of chance constraints. In: Calafiore G, Dabbene F (eds) Probabilistic and randomized methods for design under uncertainty, Springer, Heidelberg, pp 3–47

[NZ00] Neumann K, Zimmermann J (2000) Procedures for resource leveling and net present value problems in project scheduling with general temporal and resource constraints. Eur J Oper Res 127(2):425–443

[NSZ03] Neumann K, Schwindt C, Zimmermann J (2003) Project scheduling with time windows and scarce resources. Springer, Heidelberg

[OZ07] Ordóñez F, Zhao J (2007) Robust capacity expansion of network flows. Networks 50(2):136–145

[Ö98] Özdamar L (1998) On scheduling project activities with variable expenditure rates. IIE Trans 30(8):695–704

[OD97] Özdamar L, Dündar H (1997) A flexible heuristic for a multi-mode capital constrained project scheduling problem with probabilistic cash inflows. Comput Oper Res 24(12):1187–1200

[OEH02] O'Sullivan J, Edmond D, ter Hofstede AHM (2002) What's in a service? Towards accurate description of non-functional service properties. Distributed Parallel Databases 12(2–3):117–133

[Pap03] Papazoglou MP (2003) Service-oriented computing: concepts, characteristics and directions. In: 4th international conference on web information systems engineering (WISE 2003), IEEE Computer Society, Rome, Italy, pp 3–12

[Pee05] Peer J (2005) Web service composition as AI planning—a survey. Technical Report, University of St. Gallen

[PK05] Pennanen T, Koivu M (2005) Epi-convergent discretizations of stochastic programs via integration quadratures. Numer Math 100(1):141–163

[Pet75] Petrov VV (1975) Sums of independent random variables. Springer, Heidelberg

[Pfl01] Pflug GC (2001) Scenario tree generation for multiperiod financial optimization by optimal discretization. Math Program 89(2):251–271

[PW07] Pflug GC, Wozabal D (2007) Ambiguity in portfolio selection. Quant Finance 7(4):435–442

[Pin08] Pinedo ML (2008) Scheduling: theory, algorithms, and systems, 3rd edn. Springer, Heidelberg

[Pow07] Powell WB (2007) Approximate dynamic programming: solving the curses of dimensionality, Wiley, New York

[Pré95] Prékopa A (1995) Stochastic programming. Kluwer Academic, Dordrecht

[Pri66] Pritsker AAB (1966) GERT: graphical evaluation and review technique. Memorandum RM–49730–NASA, The RAND Corporation

[Put94] Puterman ML (1994) Markov decision processes: discrete stochastic dynamic programming. Wiley, New York

[RR02] Rachev ST, Römisch W (2002) Quantitative stability in stochastic programming: the method of probability metrics. Math Oper Research 27(4):792–818

[RS04] Rao J, Su X (2004) A survey of automated web service composition methods. Proceedings of the 1st international workshop on sematic web services and web process composition (SWSWPC 2004), San Diego, California, Springer, Heidelberg, pp 43–54

[RR98] Reemtsen R, Rückmann J-J (eds) (1998) Semi-infinite programming. Kluwer Academic, Dordrecht

[RU00] Rockafellar RT, Uryasev S (2000) Optimization of conditional value-at-risk. J Risk 2(3):21–41

[RBHJ07] Rosario S, Benveniste A, Haar S, Jard C (2007) Probabilistic QoS and soft contracts for transaction based web services. In: Proceeding of the 5th IEEE international conference on web services (ICWS 2007), Salt Lake City, Utah, IEEE Computer Society, pp 127–133

[Rus70] Russell AH (1970) Cash flows in networks. Manag Sci 16(5):357–373

[RS03] Ruszczyński A, Shapiro A (eds) (2003) Stochastic programming. Elsevier, Amsterdam

[Sah04] Sahinidis NV (2004) Optimization under uncertainty: state-of-the-art and opportunities. Comput Chem Eng 28(6–7):971–983

[Sch95] Schoemaker PJH (1995) Scenario planning: a tool for strategic thinking. Sloan Manag Rev 36(2):25–40

[Sch01] Scholl A (2001) Robuste Planung und Optimierung – Grundlagen, Konzepte und Methoden, Experimentelle Untersuchungen. Physica, Würzburg

[Sch05] Schwindt C (2005) Resource allocation in project management. Springer, Heidelberg

[SZ01] Schwindt C, Zimmermann J (2001) A steepest ascent approach to maximizing the net present value of projects. Math Meth Oper Res 53(3):435–450

[Sha03] Shapiro A (2003) Inference of statistical bounds for multistage stochastic programming problems. Math Meth Oper Res 58(1):57–68

[SN05] Shapiro A, Nemirovski A (2005) On complexity of stochastic programming problems. In: Jeyakumar V, Rubinov A (eds) Continuous optimization—current trends and modern applications, Springer, Hiedelberg, pp 111–146

[SNLS05] Singh J, Nookala V, Luo Z-Q, Sapatnekar S (2005) Robust gate sizing by geometric programming. In: Proceedings of the 42nd annual conference on design automation (DAC 2005), Anaheim, California, pp 315–320

[SSH99] Sutherland IE, Sproull RF, Harris DF (1999) Logical effort: designing fast CMOS circuits. Morgan Kaufmann, Los Altos

[Sti09] Stiller S (2009) Extending concepts of reliability. Network creation games, real-time scheduling, and robust optimization. PhD thesis, Technische Universität Berlin

[Tah97] Taha HA (1997) Operations research: an introduction, 6th ed., Prentice Hall, Englewood Cliffs

[TFC98] Tavares LV, Ferreira JAA, Coelho JS (1998) On the optimal management of project risk. Eur J Oper Res 107(2):451–469

[TSS06] Tilson V, Sobel MJ, Szmerekovsky JG (2006) Scheduling projects with stochastic activity duration to maximize EPV. Technical Memorandum No. 812, Case Western Reserve University

[vdA03] van der Aalst WMP (2003) Don't go with the flow: web services composition standards exposed. IEEE Intell Syst 18(1):72–76

[vdAtHKB03] van der Aalst WMP, ter Hofstede AHM, Kiepuszewski B, Barros AP (2003) Workflow patterns. Distributed Parallel Databases 14(1):5–51

[WA08] Wang W, Ahmed S (2008) Sample average approximation of expected value constrained stochastic programs. Oper Res Lett 36(5):515–519

[WWS00] Wang J, Wu Z, Sun Y (2000) Optimum activity delay in stochastic activity networks. In: Proceedings of the 7th international workshop on project management and scheduling. Osnabrück, Germany

[WGH04] Wiesinger C, Giczi D, Hochreiter R (2004) An open grid service environment for large-scale computational finance modeling systems. In: Proceedings of the international conference on computational science (ICCS 2004), Kraków, Poland, Springer, Heidelberg, pp 83–90

[WHK08] Wiesemann W, Hochreiter R, Kuhn D (2008) A stochastic programming approach for QoS-aware service composition. In: Proceedings of the 8th IEEE international symposium on cluster computing and the grid. Lyon, France

[WKR10] Wiesemann W, Kuhn D, Rustem B (2010) Maximizing the net present value of a project under uncertainty. Eur J Oper Res 202(2):356–367

[WKRa] Wiesemann W, Kuhn D, Rustem B Multi-resource allocation in stochastic project scheduling. Ann Oper Res, in press

[WKRb] Wiesemann W, Kuhn D, Rustem B Robust resource allocations in temporal networks. Math Program, available via Online First

[Wil99] Williams HP (1999) Model building in mathematical programming. Wiley, New York

[Wol85] Wollmer RD (1985) Critical path planning under uncertainty. Math Program Study 25(1):164–171

[WY06] Wu J, Yang F (2006) QoS prediction for composite web services with transactions. In: Proceedings of the 4th international conference on service-oriented computing (ICSOC 2006), Chicago, Illinois, Springer, Heidelberg, pp 86–94

[Yan05] Yang I-T (2005) Impact of budget uncertainty on project time-cost tradeoff. IEEE Trans Eng Manag 52(2):176–174

[ZBN+04] Zeng L, Benatallah B, Ngu AHH, Dumas M, Kalagnanam J, Chang H (2004) QoS-Aware middleware for web services composition. IEEE Trans Software Eng 30(5):311–327

[ZF09] Zhu S-S, Fukushima M (2009) Worst-case conditional value-at-risk with application to robust portfolio management. Oper Res 57(5):1155–1168

Index

W. Wiesemann, *Optimization of Temporal Networks under Uncertainty*,
Advances in Computational Management Science 10, DOI 10.1007/978-3-642-23427-9,

MIX
Papier aus verantwortungsvollen Quellen
Paper from responsible sources
FSC® C105338

Printed by Books on Demand, Germany